John McCarthy :

Gunship Pilot Vietnam

Romar Press

Romar Press

3157 CR 518, Stephenville, Texas 76401
publisher@romarpress.com
www.romarpress.com

Text and Illustrations Copyright © 2024 by Romar Press

Hardover: ISBN 979-8-9853609-6-7
eBook: 979-8-9853608-8-1

DEDICATION

First, I want to recognize John B. McCarthy, CW3 (REG), my father. Without his style of parenting, I would never have been able to do what I have done in my life. As a father, he taught me about being open and honest. He taught me to be all that I can be. He taught me to be an honest and sincere person who can be trusted. He taught me to stand up for what I believe is right and to defend those who can't thrive or protect themselves. I learned to do the right thing from him.

CW3 John M. McCarthy (left) and CW3 John B. McCarthy (right)

PREFACE

For Cindy my wife, I can say that I have seen her move forward in her life to be the best nurse that she can be. Growing up in western Pennsylvania, she was the first member in her entire family to attend college, earning a Bachelor's degree in science in Nursing and a Master's degree in Health Service Administration. She followed those degrees with an MBA and a Doctorate of Nursing Practice. She earned the last two degrees as we traveled and while the effects of agent orange exposure caused me to start the long road to seriously declining health. She has been the person so many people wish they could be. She is an excellent nurse and a model Chief Nursing Officer. She has shown me love and caring as a wife while being by my side through this journey of twenty surgeries and my constantly deteriorating health. She always has a smile, and she listens to my concerns regarding whatever I am dealing with.

I am truly blessed to be with these two people who have both kept me going in the right direction.

JOHN STEINBECK

John Steinbeck went to Vietnam for a year in 1966 at the request of Harry F. Guggenheim, publisher of *Newsday*. Asked to write a series of Vietnam war reports, he chose the epistolary form, addressing the letters to his dear friend Alicia Patterson, *Newsday*'s first editor and publisher.

. . . I wish I could tell you about these pilots. They make me sick with envy. They ride their vehicles the way a man controls a fine, well-trained quarter horse. They weave along stream beds, rise like swallows to clear trees, they turn and twist and dip like swifts in the evening. I watch their hands and feet on the controls, the delicacy of the coordination reminds me of the sure and seeming slow hands of (Pablo) Casals on the cello. They are truly musicians' hands and they play their controls like music and they dance them like ballerinas and they make me jealous because I want so much to do it. Remember your child night dream of perfect flight free and wonderful? It's like that, and sadly I know I never can. My hands are too old and forgetful to take orders from the command center, which speaks of updrafts and side winds, of drift and shift, or ground fire indicated by a tiny puff or flash, or a hit and all these commands must be obeyed by the musicians' hands instantly and automatically. I must take my longing out in admiration and the joy of seeing it.
January 7, 1967

Steinbeck in Vietnam: Dispatches from the War
Edited by Thomas E. Barden

In 1775, the Second Continental Congress established the Army to protect the freedom of the first thirteen colonies. Today the United States Army, the world's premier fighting force, is the most powerful military of all times. The high standards of its commanders and fighters have been codified in the *Warrior's Code* and the *Warrior's Ethos*. These beliefs demonstrate the sustaining premise of what it means to be a soldier, one who gives all, all required for the cause. Many versions of these statements exist, but none is ascribed to a single author. As a nineteen-year-old gunship pilot in the jungles of Vietnam, the following ideals kept me going and solidified my belief in our mission, even as support eroded at home. I offer no excuses for the stories I share in this book. I believed in our power to right the wrong being committed against the South Vietnamese people. We stepped up when we were asked, and we gave it all we had.

PREFACE

A Warrior has been defined as a person engaged or experienced in warfare. The warrior spirit refers to the quality within to live with *humility, passion, and courage from a place of empowerment.* A warrior's attitude means *you will stand for what you believe in.* You know the difference between *right and wrong* **and you have the courage to choose right.** A true warrior's ethos underpins the Army's enduring traditions and values and means a commitment to excellence. Every soldier must know the *Warrior's Code* and live the *Warrior's Ethos.* There are four tenets of the *Soldier's Creed* that are the basis for the warrior's ethos. He lives and fights by them:

1. I will always put the mission first.
2. I will never accept defeat.
3. I will never quit.
4. I will never leave a fallen comrade.

I am disciplined, physically and mentally tough, trained, and in my warrior's tasks and drills. I will always maintain my arms, my equipment, and myself. I am an expert, and I am professional.

This is a complex and demanding creed. These tenets may sound simple to learn, understand, and be satisfying, but when you live these rules, you see success. They have been an intricate part of the Army since the beginning. The point of the Creed has been strict and has not always been called the Creed, but it has been the basis for the behavior of Army soldiers for a very long time. It has been an integral part of the Army and lives in the hearts of soldiers and ex-soldiers from all over the United States.

Not everyone agreed with this war. As it continued, from my perspective, we should not abandon the South Vietnamese and let the Communists roll over them. A war is a very complex and ever-changing set of circumstances that may be valuable situationally while not as valuable in the overall war. In many cases of anti-war protesting, there was a permanent use of loud versus accurate.

This is my story, and my truth to tell.

TABLE OF CONTENTS

July 1971

August 1971

September 1971

October 1971

THE BEGINNING

In starting this book, I felt I had to explain why someone would want to become an Army helicopter pilot. In many ways it isn't hard to understand, but once you really interrogate the question, you find that many of the helicopter pilot trainees don't necessarily know why. For some it's because of the human desire to fly. Period. Perhaps the desire to fly was instilled as a dream in humans long before it was even vaguely possible to fly. For some, the idea of flying instead of walking into battle seemed very appealing. For me, it was a desire I had since I was a little kid.

My father was an Army helicopter pilot; I grew up around these machines and the men who flew them. When I was a young boy, helicopters had normal reciprocating engines. They were loud and powerful and took large aircraft into the air, so they could fly away. From my vantage point on the ground, the pilots fulfilled Da Vinci's dream of humans flying. In some cases, the engines were big round motors that shot flames and belched smoke out of the exhaust as they were starting. All of this with smoke and flames and belching motors had me to the point of certainty that I would be that pilot in the cockpit with the open window, watching the start sequence and monitoring the start until the clutch caught up and had the rotor blades turning. Very slowly at first, and then in just a few seconds, the blades begin to pick up speed. Then the swishing noise of the blades settles into a continuous pattern, and the start sequence is over, and the ground idle happens and would soon be the focus until the engines were up to speed. The mighty blades are at their operating speed. The pilots do their run-up checks to make sure all systems are working together, and the aircraft starts moving into take-off position.

All it took was one helicopter to grab a boy's attention and hold it until the aircraft was out of site. When several of these machines worked together, it was mesmerizing.

I was lucky enough that my family moved to Fort Churchill, Manitoba, Canada. Churchill is north of the town of Thompson which is 250 miles away. Polar bears congregate there, and we watched them when the winter was over. The bears rode the ice when it broke off into the bay where they could catch seals and walruses to eat. They played with the baby bears and taught them how to hunt these food staples, so they could stay alive.

I had been fascinated when we were living in Canada, not only with the polar bears, but also, with the helicopters. One mission my dad was involved in was flying out to retrieve the nose cone with the monkeys that were shot up in rockets to decide if piloted space flight was, in fact, practical. Everything about these aircraft was almost magical to a kid. My fascination with things in the air in this magical place was developing day by day.

This tour in Canada was a hardship tour when a soldier was unaccompanied by family, but we went with my father.

We went up to Hudson Bay, which extended my father's tour. We were a close family, so it never qualified for a hardship tour to us. We stayed there for over three years, and our family thrived.

That was not true for every family. I remember one night when the lady living next door to us had what was possibly a mental breakdown. I could hear her screaming and throwing things in the house. As I remember it, they went back to the United States where her husband could resume his job in the military, and she could get psychological help. The isolation of living that far from any other town was more than some could stand.

The very cold, minus fifty-degree temperatures required a great deal of time and stamina. Snow up to the windows of a two-story building made leaving our quarters a rare event in the worst part of winter. At the time, this was the place I had lived the longest of my life. The helicopters could fly in some of this terrible weather but not when it was so cold. I remember learning that in some weather conditions, even a helicopter cannot fly, but this knowledge did not make me abandon my dream of flying in the Army.

We went on to Dad's next assignment which I believe was to Fort Belvoir, Virginia. After that, we went to Germany, which I absolutely loved. Outside the US, the Army always had activities and services for dependent children, and I was especially interested in sports. I played in most sports where we would travel by bus to other Army posts, most of which had far above average facilities. The aviation assets in Germany were strong, and at many of these posts there were airfields with helicopters and Army airplanes. All the impressions I gained strengthened my resolve to become an Army aviator.

At that time as I was becoming a teenager, I had no idea what I had to do to get into flight school. I didn't worry about failing *out* of flight school because it made sense to me that I had to be admitted *into* flight school before worrying about busting out. My father had fought in Korea in the Korean War. He had not gone to flight school before the Korean War, so he was a regular line soldier. I somehow felt like he would not have to fight in the Vietnam War since he had already been in one brutal war during his career. The truth is that I had heard little about Vietnam, and at this point did not even know where it was. As uninformed as I was, it was a complete shock when one day I came home from school, and my father was home. My first thoughts were that I must have done something wrong that brought him home to punish me. When my sister arrived home, my father told us that we had to return to school to withdraw from all our classes and then come home to pack what we would need for about a week. I don't remember how we got back to school because it was about forty-five minutes away. I was sure the teachers had already left for the day, but when we arrived, I was told our teachers would be ready with the withdrawal paperwork. However, the Army people who set this up didn't understand our timeline. Our final exams were scheduled in a week or two. No one counted on the fact that one of my teachers would *not* release me until I took the final exam. Things would certainly be on this exam that the teacher would cover next week, so I had no clue what was on that test. I had fair grades, but I knew if I asked for allowances, this teacher would not grant them. We were supposed to leave the next day, but because of my circumstances, we were there for two more weeks staying above the Officers Club in an apartment made available for emergency purposes. I did *not* fail the exam.

I saw my friends before and after school and took the friendly harassment as it was intended. By now there were other students going through the same thing as I was with our fathers going to war.

We then settled in California at an investment house my uncle owned. Shortly after our arrival, my father headed to Vietnam to join the First Cavalry. This came at a significant moment in the war effort because for the first time in history, helicopters were used as an integral part of our forces, particularly at the Battle of la Drang Valley. This would be our military's indoctrination to this kind of warfare. Interestingly, many years later, this fact caught the attention of filmmakers, and the events surrounding the First Cavalry during this time are chronicled in the movie *We Were Soldiers*, starring Mel Gibson.

I know everything about what it means to be at home when a loved one is fighting in a war. I feel for my mother who went through all of this with her husband and then with me when I enlisted and went to Vietnam.

It was tough being at home while my father was overseas fighting in the war against Ho Chi Minh's invasion of South Vietnam, a sovereign nation that preferred to stay free of Communist rule. We watched the news every night, and seeing and hearing the news reports were unnerving to say the least. The letters and audio tapes that we received from my father told of a war that was at odds with the news reports. I was only fifteen during this time of unrest in our country, and these differing reports caused a great deal of confusion for me.

My father was always my hero, and I did not believe he was lying to us, and yet we were taught to trust the news reports, which differed from his stories home. Vietnam has been called the first war televised. Journalists were reporting their perspectives from the field, and their unfiltered reports were often at odds with the soldiers' realities. These differences led to the chasm created in our country and to my general mistrust of mainstream media. It was when I was in the war in Vietnam that I realized my father's version of the war was much closer to reality than what was playing out on television.

The first engagement of helicopters in a full-scale attack became an engraved part of the story of my young life. My father was in the Battle of the la Drang Valley, the first major engagement of regular U.S. forces against the North Vietnamese

Army. In the first four days of this campaign 234 Americans were killed and 250 were wounded. In the 43-day campaign 545 Americans were killed. Enemy deaths have been estimated at 3,561 with no number of wounded, but the numbers were unbelievably high.

I did not set out to be a hero, and I was not one. I was just a young man doing my job to the best of my ability. Being a dependent with a father serving in Vietnam meant that I met and knew well many other Army kids. Growing up together in this small circle of Army life gave me a glimpse of true loss. Some of those kids that I grew up with had fathers who were some of the many helicopter crews lost in the Vietnam War. They were real heroes. These pilots set the bar extremely high for those of us who followed them. They showed the world what Army aviators can and will do.

From these remarkable pilots came the standards we would all follow. We learned that you must put the mission above all else regardless of the enemy size and armament. We would not accept defeat. In my experience, it did not matter what the enemy was throwing at us or how many of them there were. My job was to turn inbound and fight with all the weapons I had.

We covered helicopters, both Medevac and slicks in the area with whatever we had. During some of these events, we were running low on ammo, but the enemy did not know that, so we would use our remaining ammo judicially. Doing this gave them a few extra shots at us, but we knew we were on the side of right, and we were able to defeat them. We knew that fighting was the right thing to do. These were brutal invaders, as were those who betrayed their families to become Communist fighters. We fought to the death.

John McCarthy :

Gunship Pilot Vietnam

INDIVIDUAL FLIGHT RECORD 1971: MAR - JUN

INDIVIDUAL FLIGHT RECORD . J FLIGHT CERTIFICATE - ARMY — 26. PE..ID COVERED — 27. SHEET NUMBER
(PART II)
For use of this form, see AR 95-64, the proponent agency is Office of the Assistant Chief of Staff for Force Development

1971: Mar - Jun — **2-1**

28. LAST NAME - FIRST NAME - MIDDLE INITIAL: **MCCARTHY, JOHN M.**
29. SERVICE NUMBER 69AN
30. GRADE AND COMPONENT: **WO1 USAR**

SECTION V - FLIGHT HOURS ACCRUED - TOTAL HOURS FLOWN BY MONTH *(For Local Use as Desired)*

JULY	AUGUST	SEPTEMBER	OCTOBER	NOVEMBER	DECEMBER	JANUARY	FEBRUARY	MARCH	APRIL	MAY	JUNE

SECTION VI - RECORD OF FLYING TIME

					FIRST PILOT FLYING TIME								COPILOT FLYING TIME						
						FIXED WING				ROTARY WING					FIXED WING			ROTARY WING		
							NIGHT				NIGHT					NIGHT			NIGHT	
DATE	AIRCRAFT TYPE MODEL SERIES	MISSION SYMBOL	AIRCRAFT COMMANDER	INSTRUCTOR PILOT	FIRST PILOT	WX INST	VFR	WX INST	HOOD	WX INST	VFR	WX INST	HOOD	CO PILOT	WX INST	VFR	X INST	WX INST	VFR	WX INST	CROSS COUNTRY
a	b	c	d	e	f	g	h	i	j	k	l	m	n	o	p	q	r	s	t	u	v
March																					
NO TIME FLOWN DURING THIS PERIOD																					
April																					
30	UH-1H	C			2.3																
May																					
2	UH-1H	C			11.6						1.9										
3	UH-1H	C			8.8						1.0			1.3							N
5	UH-1C	C			0.8																N
6	UH-1C	C			2.5																
7	UH-1C	C			6.5																X
9	UH-1C	C			2.5																X
10	UH-1C	C			2.5																X
14	UH-1C	C			6.1																X
17	UH-1C	C			2.8																X
19	UH-1C	C			2.8																X
22	UH-1C	C			3.5							1.0									H
23	UH-1H	C			4.8						1.0										N
24	UH-1C	C			3.0																X
25	UH-1H	C			2.0																X
27	UH-1C	C			7.5							0.5									N
28	UH-1C	C			5.5						2.0										N
29	UH-1C	C			4.3																X
30	UH-1H	C			4.3																X
31	UH-1C	C			1.8																
					/////LAST ENTRY/////																

31. TOTALS THIS SHEET					86						6	1	1	1							
32. TOTAL BROUGHT FORWARD FROM SHEET NO. 1-1					9						↑										
33. TOTALS TO DATE					95						7	1	1	1							

DA FORM 759-1 (PART II)

FLIGHT SCHOOL

E very endeavor must have a starting point, and for me, it was at the U.S. Army Primary Aviation Center at Fort Wolters, located in the small town of Mineral Wells, Texas. Residents still talk about those years when helicopters filled the skies, making the very air vibrate with the sounds of the humming blades. The year was 1970, and the flight school not only trained these men (no female aviators in 1970) to be professional aviators, but it also trained us to be professional officers.

Simply put, I had gone from a high school graduate at the age of seventeen to starting college, turning eighteen, knowing I would only be there until I could take and pass the "Fast Test." This Army test indicates whether or not you could be a candidate for flight school. It can only be taken once, so the rule is pass or forget my dream of being an Army helicopter pilot. The test was many hours long with many parts. Each part was timed, and the time was strictly enforced. The questions covered many items, some of which seemed to have no relation to flying or to the Army, for that matter. Four of us took the test on the day that I did, and only two of us completed it.

I had completed all the entrance requirements before entering the Army, including testing and passing a flight physical, so I had a date and class number before entering the Army. I went to Fort Polk, Louisiana, for basic training. Then I returned to Mineral Wells to enter flight school at Fort Wolters, Texas.

Flight school was roughly ten months. Those ten months were long, hard months. The intense program combined learning how to fly helicopters and how to become an Army officer.

My day usually started at around 0500 and ended at 2200 hours. After this seventeen-hour day was over, my nighttime activity started. After bed check, we would generally do the cleaning that was required. The concrete floors were waxed and buffed until they had a mirror-like surface with no smudges. The bathrooms were spotless. Our gear was polished, shined, and folded exactly as described in the appropriate manuals. Everything in my area had to be shining and perfectly folded, underwear included with folds measuring exactly six inches. Every millimeter of my brass belt had to be polished and shined with no scratches.

In the event things weren't as perfect as expected, we were handed demerits. It took only a few demerits to keep a candidate on post over the weekend with more cleaning, scrubbing, and folding to do. Demerits for military failures meant "taxi time." The offending candidate would be required to march on an area about the size of two basketball courts that had straight and squared lines. The TAC officers determined the appropriate time length assigned for this activity.

The honor code was the same as West Point's Honor Code, and the best way to describe it is "Draconian." Should a candidate lie and be aware he had lied, he had to report it or be in violation. This circumstance proved to be grounds for elimination, and to the best of my knowledge, it was enforced. The process involved a strong degree of honesty and integrity, and it weeded out people who refused to give up everything to make it through this program.

The numbers of candidates who failed were high. Some soldiers self-eliminated because they couldn't take the pressure for one more day. At the time, many of us candidates didn't understand the reasons for such a stringent process that weeded so many out, but when I arrived in Vietnam, I began to understand. The use of helicopters in war time the way they were being used in this war was relatively new. This would be a testing ground, and helicopters were proving to be an amazing asset to the military. Using helicopters to medevac wounded soldiers and speed troops into battle were likely beyond the imagination of what helicopter proponents envisioned. By virtue of the visibility, the helicopter brought the battle command and control by senior officers into real time.

I had one big advantage: my father. When I was in the tenth grade, my father went to Vietnam with the 1st Cav. He was in the IA Drang Valley in that very violent part of the war. He sent us audio tapes about once a week and letters very often. I soon concluded that most people knew nothing at all about the reality of the Vietnam War. Those who relied only on the media with no communication from soldiers who were there, knew less than nothing. I had no intention of riding on my father's successes, so I didn't often tell people that I was following in *his* footsteps – literally – experiencing the same war, flying the same kinds of helicopters, fighting the same kinds of battles. I did not want other pilots to think that I got breaks because of him. I was, and still am, extremely proud of my father as a man, a father, and an Army Aviator. I decided, however, that I would not make it because of him. If I made it, it would be because of me.

HELICOPTERS ARE COMPLICATED MACHINES

I n flight training, certain things must occur by a certain time, or you could, and probably would, be eliminated from the program. Being eliminated can have a huge effect on a person. Most candidates work hard to get into flight school and then work even harder to stay in. Being eliminated causes a person to look at what their failures are. *Why wasn't I as good as the other candidates who are progressing with no apparent problems? Was my elimination truly my fault, or did one of the TAC officers have it in for me? How will I act if I run into any of my classmates who are now Army pilots and Warrant Officers while I am a Spec4, a situation that would mean they can be my boss, and I cannot mingle with them?*

Many candidates were waiting to get married when they graduated, and for those who were eliminated, the loss was staggering; they lost the pay raise that is part of the higher rank and the prestige of being a U.S. Army Aviator. In some cases, candidates' fiancées left because the soldier hadn't recovered from the failure. In addition, most of the new guys who "busted out" went to mortar training and then to Vietnam to "hump mortars" through the jungle.

Consider this: you have trained as an infantry soldier, ready to deploy as a grunt. Many things drive an individual to reach down and grab another bit of the "I will not be washed out. I will power through whatever the issue is." You study harder and better and apply yourself in the daily physical training. You must be "all in" to make it through this course. The statistics are against you; almost half of the candidates will not make it through. I was in the half that made it.

It is not necessarily only a failure on the candidate's part that can cause dismissal from the program. Candidates know that machines can and do fail, a fact made clear by the inordinate amount of time spent practicing failure recovery. Fortunately, a helicopter is very manageable in the event of engine failure IF the conditions are right: it's a beautiful day with low winds, not excessively hot, and you're in an open field. And you've practiced the maneuver many, many times. It's plausible for the engine to quit when it wants to.

The Army Flight Program requires three complete solo trips in three days around the landing zone in three days. All went well on my first solo flight of three touchdowns, and the same on my second day set of three. On the third day my luck changed.

I taxi down to the takeoff point, do my pre-takeoff checklist. The tower clears me, so I move the cyclic forward, gaining airspeed and altitude slowly and smoothly with about the same amount of power needed to hover the aircraft. I gain about twenty feet of altitude and move about ten feet forward when I notice a huge pile of cactus ahead of me.

Then "boom," I hear a loud explosion, and the cyclic almost jerks out of my hand. I pull it aft, hoping to land on the pad I had departed from and not on that cactus. I see that I'm about to hit the ground, and I have full pitch, which can cause a rapid climb, but with a dead engine, it just increases the angle of attack on the blades. It cushions my touchdown, and somehow, I'm back on the pad I had just left from, completing a successful backwards autorotation.

As I emerge from the aircraft everyone heads over to see me and find out what happened. The engine had failed catastrophically. One of the cylinders had blown the head of the piston cover completely off, and the piston was hanging out.

"No skill involved," I told everyone. "I just made movements in reaction to what was happening." But I was scared. Scared that I would be eliminated from the program. I had only a few minutes over a meager ten flight hours, and I figured inspections would show something I had done wrong. Luckily, the pilot in the aircraft behind me explained, "I heard and saw the whole thing. He didn't add a bunch of throttle to increase his rate of climb. That engine failed catastrophically."

The next day, the review board calls me in. "Congratulations on your excellent recovery," they announced. "You showed no signs of fear to continue flying." And fear didn't plague me, then or ever. When I went to Vietnam or later after the Army in private enterprise, fear didn't infect my flying. This solo experience began my thirty-one years of flying helicopters, and it taught me to never quit flying and to keep my head in the game. Knowing what to do in the event of aircraft failure proved to me that I could do this job and do it right. I would be ready to fly combat helicopters in Vietnam.

ARRIVAL

I arrived in Saigon after being squeezed between two other soldiers. (We were all FNGs - a term to describe F**** New Guys) on a series of long flights in a stretch DC 8. The aircraft had been lengthened from its original DC 8 form to add extra rows. I don't know if this was done under a government contract or not, but just as luck would have it, this was bad luck. Not only did they add rows of seats to carry more people for the military, they also closed the rows a few inches. We sat crammed together to fly from Seattle Tacoma (Sea Tac) Airport to Alaska, our first stop. We arrived in the dead middle of the night, and nothing was open that sold food or drinks, but once we boarded, fortunately, we were served box lunches and a coke to drink. All we did was walk around in the terminal for a few minutes until the aircraft was refueled, and we were told to get onboard. On we went.

From Alaska we flew to Guam where once again we stopped for the aircraft to be fueled. Again, we were served box lunches and a coke to drink. On we went.

Our next stop was in Manila in the Philippines and then on to Saigon, Vietnam, another thousand miles. By the time we arrived, we were all exhausted from lack of sleep and real food and water. It is just a guess on my part, but I believe we had all of those stops because of the increased seating and, therefore, increased weight. We used far more fuel than the aircraft would have used before the modification. This flight was roughly 7200 miles.

As we disembarked from the airplane, we were now in a country at war. I don't know what I expected, but it was not what I saw. Non-military aircraft were on the ground, and I noticed a C-130 Air Force airplane with a line of combat-ready troops boarding. These troops were all carrying loaded backpacks, front packs, and their M-16 rifles. From my vantage point, they all looked weary. I didn't know what their destiny or purpose was, but it was a wake-up call. The war was raging.

We waited in a very loose formation as instructed for about half an hour until our rides showed up. I expected Army deuce and half-trucks. Instead, there were blue buses with wire cages over all the windows. I wondered about security since only wire guarded the windows on buses with no other protection. A VC could ride up, tape a grenade to the bus, and pull the pin. In about seven seconds, the explosion would take place. No one could stop it. At least it's safe to say that nobody would launch a grenade through an open window. Any protection is better than no protection in this case.

We boarded our bus and headed for the replacement depot where all incoming and departing soldiers checked in, and paperwork was documented. The Army is meticulous about paperwork, but even more so in a combat zone.

As we made our way from the airport to the replacement depot, I was fascinated with how busy Saigon was. Swarms of people were everywhere on motorcycles, scooters, and bicycles. Cars and trucks didn't seem to follow any road rules except the ones they wanted to follow. Some of the motorcycles had as many as four people on them, some with small children. I didn't expect this much civilian movement since a very active war was not that far away. What I was witnessing was mad confusion.

We finally arrived at the repo-depot and disembarked from the bus. The commanders split up the officers and enlisted service members, and so we were in a much smaller group. We were told where to go, so we went inside as directed and picked out a bunk. The setting had the traditional décor of Army barracks without the spit-shined floors. We were told to relax, where the chow hall was, and where the Post Exchange (PX), the Army's department store, was located. A couple of us walked to the PX to see what they carried. We appreciated this chance to stretch our legs after what we thought was the flight from hell. Little did we know.

We were surprised by the size of the store and what they carried. Top-of-the-line stereo equipment, watches, and other items to make life better. Since we didn't know what we could have, our visit was short. We looked around for a few minutes, and then left.

Our next stop was a medium-sized officers club where we discovered we could order hot or cold sandwiches and beer. When I saw the Budweiser, it made me realize it was not all bad. After a couple of sandwiches and a couple of beers, we headed back to the barracks. Within minutes of arriving, we were sound asleep, my one and only good night's sleep I would have in about a year.

The next day, assignments were posted, but none for the aviators in the group. We knew of a requirement that allowed us to meet with an Aviation Assignments Officer regarding where we would be placed. That was a rule I did not see followed, but we were told that we would see our assignments and be on our way the next day. That didn't happen, but at least we were able to get rested and fed and have a few beers.

On the third day at about noon, the pilot placements started showing up. Mine came out about four o'clock in the afternoon, and I saw I was to leave early in the AM. A few of my flight school classmates were there at the same time, and it helped me not to feel so out of place. Finally, the next morning we went to the loading point and boarded the buses to head to the airport for departure up country. All the pilots on this trip would all be in II (two) Corp., otherwise known as the Central Highlands. We would not all be in the same company, but we would possibly see each other somewhere down the road and catch up.

We arrived at the airport with no problems and headed over to the C-130 that would move us closer to our assignments. As we were moved upcountry to our assigned units, someone from that unit met us. As we narrowed down to only three FNG pilots on this C-130, as well as quite a few grunts, we landed at Qui Nhon to meet our company representatives.

The load master on this C-130 screamed at us at the top of his voice, "Get off. Get off the plane. Move it. Get off NOW!"

I had been yelled at for over a year during basic training by drill sergeants, and now by the Load Master in front of me who was still screaming. I tactfully explained how the military rank structure worked, "You're going to have to get someone with more horsepower to make the three of us move any faster."

The Load Master took off and reappeared with a bird Colonel, who was the Aircraft Commander (AC), so we dutifully disembarked from the aircraft quickly, as instructed by the AC. When we were off, we saw fuel pouring from the wing where we had taken a large round through it. Had the round been a few inches over it would have taken out the engine and possibly set the wing on fire. My first flight in Vietnam ended with a quick unload due to enemy fire, not a resoundingly non-stressful flight and kind of eye-opening for a young pilot.

I left that airplane with a great lesson learned. Before you draw a line in the sand to force an action, make damn sure you know the situation and who is controlling the other side of it. That enlisted man load master had the power, and we got to see it firsthand.

NIGHTTIME REFLECTIONS

Nighttime reflections brought to me my first time thinking deeply about what was happening with my going to Vietnam and about what I would become. Looking now at the world through the eyes of a combat aviator, I knew I had changed, and not just a little bit, because of my combat experiences. As far as where I am today, it'd take more than this book to analyze me. I can't say that I was ever scared or wanted to opt out of the tough decisions that come with those special flights. My thoughts about where I was weren't fear based, but I managed to live for the challenges, so I was all in. Sometimes we knew it was going to be tough. We were involved in ammunition monitoring due to large bouts with miniguns and M-60's. Sometimes we would have to face the fact that we used up our rockets.

On a mission which is basically "go," we had reports of a large number of enemy troops that just arrived there, so obviously we were preparing for a big battle; however, I learned not count on those reports, that whatever information came in, the powers that be probably had things back-asswards, so I became a skeptic. I recognized in myself that I had a great deal of mistrust now. I did not before.

Having learned how to become more situationally aware also alleviated fear. If you stay on guard, you don't have fear. That might be for no other reason than you were too busy for fear to take over some of your thinking. Power to push through the fear came from having the knowledge and understanding that things could go all to hell in seconds. Knowledge and understanding don't contribute to being scared. This way of thinking started changing my whole way of looking at life.

I worked with the awareness that the end could happen to me just as it could to any other pilots. I was able to put that fact to rest flying, but when the battle was over and my flying day was done, I could not help but review everything I had been through. It wasn't uncommon to notice a little shaking in my hands and sweat breaking out on my forehead as I relived the events. Scared, maybe I was, but I believe it was with the knowledge that the littlest thing could have altered the results. For example, I was flying out of trim, a situation in which the aircraft isn't moving through the airmass smoothly. We were fired on, and the round missed us. If we had been in trim, we would have taken the hit. Near-misses are made out of things that might look inconsequential but constitute important things. I realized early nothing positive would come from dwelling on things that did not happen. Thinking about how close we were to being killed did none of us any good.

My biggest problem lying down in the dark was thinking about war in general and how it impacted my religious outlook. I was an altar boy, now being called altar servers, and was a very strong believer in what the Catholic Church said. As I grew up, I discovered that none of us can live totally true to our religious demands. I believe that most people have flaws, some more than others. I learned to make peace with conflicting ideas and decided we can only do our best to live on the side of goodness. At night in my bunk, I felt totally removed from my desire to live to my highest goals. I usually ended these brain gymnastics by making a pact with myself, and my God, that I will do the best I can and hope he will forgive my weaknesses and failings. As hard as it may be to believe, when I thought I might be able to sleep, I tried to go over aircraft limitations and emergency procedures. This practice helped ensure that these vital procedures and details would come to the front of my brain whenever I needed that info in the heat of battle.

Often, I would think about my friends and former girlfriend and imagine what they may have been doing at the time I was running though procedures in my head in the dark of night. I never wrote to any of them. I had seen the consequences of what happened to a pilot and his young wife of about twenty years old. She was actually one of my classmates in school, married a pilot, and then became a young widow.

As strange as it sounds, she would later meet and marry *another* pilot who received orders to go to Vietnam. Unfortunately, this second husband was also killed. She went into shock seeing her life fall apart a second time because of a tragedy that occurred in Vietnam. I ran into her one day when I was home on leave after my tour, and she literally had a fit. She screamed at me, "All of you young pilots have a death wish."

"No," I replied. "I have a life wish."

I made an effort and asked why we shouldn't, the strongest country in the world, use our capabilities to stop murders, rape, and carnage, and send the invaders back to their homes or into the ground where many of us believe they belonged. Since we have the power to help the people of South Vietnam, why shouldn't we use it?

Her answer was basic. She told me I was not smart enough to figure it out.

LESSONS LEARNED

1. It doesn't matter how big an aircraft or how fast I can fly. If enough of these enemy troops fire at us, and even if only one round hits us, it can be a bad situation. Had that round entered the wing just a split second later there's a good possibility it would have taken that engine out, or worse, the fuel may have caught fire, and we would've been flaming all the way to the airport. I knew I had to remember the elevation and/or speed could not guarantee they couldn't hit the aircraft and take us down well short of our destination, plunging us into an entirely hostile situation.

2. The enemy is possibly just as dedicated to their outcome as we are. Never downplay their bravery or desire. They may be more primitive in appearance but these citizen soldiers and the North Vietnamese soldiers have been fighting for their entire lives. Vietnam had been an invaded country by

the Japanese, the Chinese and many other nations. These people were well versed in their life at war.

3. Fighting a war where body count is the judgement of success seems less than of an effective way of determining winners or losers, but body count has always been a factor, Determining what resources are needed often predicts the number of enemy forces we may face. The idea of it being the only tool to judge success by is useless if I don't know how many people they are willing to lose.

4. Fighting a guerilla-style war is far more difficult than fighting a war over property. After all, America was won by using many guerilla-style tactics.

Screened shot of the nose cone from John's aircraft. Why "Rebel"? A rebel is a person who resists or defies rules or norms, one who rises up against the powers that be—in this case Ho Chi Minh and his fighters who invaded South Vietnam with the clearly stated intention of making the South and North Vietnam one country under his rule.

INDIVIDUAL FLIGHT RECORD 1971: MAR - JUN

INDIVIDUAL FLIGHT RECORD,) FLIGHT CERTIFICATE - ARMY (PART II)	26. P\. ID COVERED	27. SHEET NUMBER
For use of this form, see AR 95-64, the proponent agency is Office of the Aviation Chief of Staff for Force Development	1971: Mar – Jun	2-1

28. LAST NAME . FIRST NAME - MIDDLE INITIAL	29. SERVICE NUMBER SSAN	30. GRADE AND CON-PONENT
MCCARTHY, JOHN M.		WO1 USAR

SECTION V - FLIGHT HOURS ACCRUED - TOTAL HOURS FLOWN BY MONTH *(For Local Use as Desired)*

JULY	AUGUST	SEPTEMBER	OCTOBER	NOVEMBER	DECEMBER	JANUARY	FEBRUARY	MARCH	APRIL	MAY	JUNE

SECTION VI - RECORD OF FLYING TIME

Columns: a DATE · b AIRCRAFT TYPE MODEL SERIES · c MISSION SYMBOL · d AIRCRAFT COMMANDER · e INSTRUCTOR PILOT · f FIRST PILOT · g–j FIXED WING (WX INST, NIGHT VFR, NIGHT WX INST, HOOD) · k–n ROTARY WING (WX INST, NIGHT VFR, NIGHT WX INST, HOOD) · o CO PILOT · p–r COPILOT FIXED WING (WX INST, NIGHT VFR, NIGHT WX INST) · s–u COPILOT ROTARY WING (WX INST, NIGHT VFR, NIGHT WX INST) · v CROSS COUNTRY

DATE (a)	TYPE (b)	MSN (c)	FIRST PILOT (f)	RW NGT VFR (l)	RW NGT WX (m)	RW HOOD (n)	CO PILOT (o)	X-CTRY (v)
March								
NO TIME FLOWN DURING THIS PERIOD								
April								
30	UH-1H	C	2.3					
May								
2	UH-1H	C	11.6	1.9				N
3	UH-1H	C	8.8	1.0			1.3	N
5	UH-1C	C	0.8					
6	UH-1C	C	2.5					
7	UH-1C	C	6.5					
9	UH-1C	C	2.5					I
10	UH-1C	C	2.5					I
14	UH-1C	C	6.1					I
17	UH-1C	C	2.8					I
19	UH-1C	C	2.8					I
22	UH-1C	C	3.5			1.0		H
23	UH-1M	C	4.8	1.0				N
24	UH-1C	C	3.0					I
25	UH-1M	C	2.0					I
27	UH-1C	C	7.5		0.5			W
28	UH-1C	C	5.5	2.0				N
29	UH-1C	C	4.3					I
30	UH-1M	C	4.3					I
31	UH-1C	C	1.8					I
////LAST ENTRY////								

31. TOTALS THIS SHEET	86		6	1	1	1	
32. TOTAL BROUGHT FORWARD FROM SHEET NO. 1—1	9		↑				
33. TOTALS TO DATE	95		7	1	1	1	

DA FORM 759-1 (PART II) 1 MAR 60

MY FIRST GUNSHIP CHECK RIDE

After another almost sleepless night, I crawled out of my bunk to get ready for my first check ride. I had to take a UH-1H check ride, so I could do some flying in slicks before I would be a full-time gun pilot, at least that was the plan. An explanation is needed here. As a rule, most gun units (UH-1C or UH-1M) and (C=Charlie model M=Mike model) required a pilot to fly six months in UH1-H slicks to learn how the slicks operated in a combat environment. The theory was it made it a more cohesive operation, and that theory became an unwritten rule. This Commanding Officer had been around long enough to have learned to trust his observations of pilots, and he trusted his judgement enough to act on his ideas.

I volunteered for scouts since no gun slots opened in gun transitions when I graduated from flight training. We occasionally heard rumors about the numbers regarding how long a new scout pilot would survive. The number I saw most was a disconcerting nineteen minutes. Since I had absolutely no idea where these numbers came from, I paid no attention to them.

Training was efficient, effective, and brief. First, I completed the OH-6 transition course at Fort Rucker, Alabama, the home of Army Aviation. Then I finished a month-long Air Cavalry Qualification course at Fort Knox, Kentucky. After Kentucky, I had a much-appreciated two-week leave. Then I was off to Vietnam. Soon after arrival, we moved upcountry to our assigned units where we were met by an individual from that unit. We went into the battalion headquarters building to see what was next on the agenda for us. As we walked into a briefing room, we saw seats and a podium. I saw a fair number of majors standing around, and they all appeared to be sizing us up. After an introduction from the Bird Colonel, we were told why the COs were there. There had been a C.O.'s meeting and these Commanders had opted to stay around to check out the new crop of pilots.

As things were going, I had no expectation of anything working in my favor since I was assigned to the 1/9 Cavalry as a scout pilot. My life expectancy was somewhere in the first half of an hour at best. I had volunteered for the Army, Flight School, and the Cavalry so I had no right to complain. In those first few minutes, I realized I had learned a second lesson: before volunteering find out as much as possible before raising your hand to volunteer. This seemed basic enough, but it had somehow escaped me.

What happened next constituted what can only be called a "life-changing" event. It had not been planned, but here it was, about to happen. A Company Commander from the 92nd Assault Helicopter Company started talking to me. He was very pleasant, but I noticed he seemed to be on edge, his eyes taking in the room from one side to the other. Growing up in the military, I knew what was going on. This C.O. asked for some of my paperwork, and I complied by handing it over.

After a couple of minutes had passed, he had me move over with him where we could talk in private.

"John," he said, "I'm offering you a slot in my company as a slick pilot. Going to the 1/9th won't be a pretty picture."

"Sir," I began in a near-whisper, "I don't want to be a slick pilot, God bless every one of them. What I want to do is fly guns, but there weren't any classes open when I graduated.

"Ok," he interrupted. "My company has a gun platoon, and I can put you in it. You'll have to do six months in a slick platoon. Then you'll do another six months in guns until you derose." (Date Expected Return Overseas or "back to the world.")

"Well, sir," I very much appreciate the offer, but I don't want to fly slicks."

He sighed, and as he was walking away, he stopped, turned around, and looked me in the eye. "Why? Why would pass up a slot that would put you in what you want for fully half of your tour?"

As uncomfortable as his question made me, I continued, "I'm not being rude or trying to take away any of the skills we all use to fight and survive. I have a personal reason that I prefer to keep to myself if you don't mind."

I could tell that my response had thrown him a little off-guard. "Ok, John, if your reason is something personal, I'll accept that. Good luck."

As he was turning to leave again, he stopped one more time. Without any hesitation, he said, "Ok, have it your way. I'll put you in the gun platoon when we get to the company area."

I was in total shock. As far as I knew he didn't know me nor my father, a Regular Army CW-3 (Retired) Army Helicopter pilot. One of the things I worried about was me getting special treatment due to my father's excellent reputation. I greatly admired my father both as a father and for his military service in two wars and peacetime. I knew I had to make it on my own in Army Aviation, or any victory I would earn would be hollow.

As the Commanding Officer and I waited to get my orders changed, we talked. He explained what I should expect from some of the other pilots. My age would be a problem for some because I was 19 years and 8 months old. I would be taking a slot from some of slick pilots who had done their six months of slick time and were waiting for an opening in the gun platoon, so they could move to the guns. All things considered, it was going to be a rocky road. I would have to stand my ground in a very respectful way to prove that I was capable and could handle these issues. My guess is that during the conversation with the Commanding Officer, he decided in those few moments that I had the strong will of a gun pilot and that was what swayed him. This special treatment was not because he had any knowledge of my father and his Vietnam service. I had done this on my own.

Things started to move extremely fast. Immediately following my Uh1-H check ride, the Commander decided that I should do an orientation check ride in one of the UH-1C gunships. Since I had flown a Charlie model (C) in flight school, I was qualified in it as well as a D model Huey, but I needed an orientation ride in one that was an active gunship.

All things being even, I had done well enough in my H model ride, which I had also flown in flight school, that they were comfortable letting me go right into the C model check. Other than having all the weapon systems on, Once the gunship was loaded, I flew a fully loaded aircraft for the first time. With the excessive temperatures outside, the aircraft's performance was definitely degraded by weight and heat. I saw for the first time what it was to fly the aircraft at near maximum gross weight in this Vietnam oven. I had no trouble flying this aircraft, but my control touch needed work, and I would get that skill down right away. We went out and flew this heavy beast around and then went to the area used as a range when test firing was needed.

The practice was varied and intense. We fired the weapons in many ways, ranging from using rocket sights to full flexing miniguns to firing stowed rockets and from the left side with no rocket sight. I had wanted this job. I had dreamed of this job. I knew right then that I had found my calling for my time in this war.

It all went well, and that was fortunate because exceedingly early, about 0415 the next morning, I was awakened by the CQ (Charge of Quarters). I was current and qualified in the "C" model Huey from flight school and I had taken an "H" model check ride here in Vietnam. So as soon as the sun was up off, we went for a quick check ride in the loaded "C" model and then it was on to refuel and take a few minutes break for a briefing and we joined up to go to fight.

We had been added to a flight of two that was going out into what the CQ believed to be a fight leftover from the Lamson Operation. I had no time to wonder, worry, or even think about this. It was a quick way into armed fighting. We were about to participate in some serious flying. The best I could do was to put my gear together: my survival vest, my helmet, Nomex gloves, and my sidearm. I had a minute as I was gathering everything up to wish that I had chosen the .45 instead of a .38. My father had given me a silver dollar that he had carried during his Vietnam service. I was glad to have his good luck piece on this first morning. I noted the heavy activity since all the guns were going to work early this A.M.

I met up with the Aircraft Commander (AC) heading out of our quarters and to the aircraft. With little time to talk, I followed his orders and preflighted the rotor head and rotor blades as he preflighted the weapons and engine. I then went to the tail rotor when the Crew Chief came up to me and introduced himself and the gunner.

We had no time to talk. As we strapped in and set everything up, the AC started the aircraft. The other two gunships were also starting, and in what seemed no time at all, we were moving into position. We bounced off the skids a couple times to move into position because these gunships were so heavily loaded. The Team lead made the radio calls to get us going. AS we departed there were other slicks and guns getting started for their day. It was all done at what seemed to be hyper speed and my head was spinning as I tried to keep up. Once we were up and in formation heading towards our Landing Zone, we had time to talk. The AC did quick introductions and an apology for having to meet this way, but this is war, and you do what is necessary and adapt. He explained our mission which was to fly gun cover in escorting and extracting some ARVN (Army Republic of Vietnam) troops in what a major event of post Lamson fighting. It had been intense and overnight the NVA started getting the upper hand and apparently had many reserve troops to replace any that were taken out. This was a last-minute mission and thankfully I had gotten check rides and orientations out of the way yesterday because that gave the platoon a much-needed extra gunship. There was a couple of minutes of bull shitting before we had to get ready for the battle. We had test fired all of the weapons shortly after takeoff, and we were good. I remember how I felt through all of this and surprisingly it was not fear. It was going over in my head what my role was in this, and I was seriously concentrating. As we started our approach to land, we were bounced to a specific spot to cover slicks going in to extract troops. We were a team of three or a "heavy team" and the patterns were worked out and suddenly, we went to work. It was a crazy world of radios going and calls being made in barely understandable English. This area was hot, and many calls were made, and Taking Fire was heard over and over with where the fire was from added in. There were some very obviously spooked troops, and some very intense slick pilots calls that we answered and covered as well as we could. I have no idea how many rounds were fired at the helicopters, but it was a large number of small arms and machine-gun fired ammunition. We got through that first lift and suddenly we were in a pattern and working hard. Once these extractions were complete, we had to stop at a field fuel and ammo stop to rearm and refuel. We took a cigarette break on the way and then stopped for hot rearming and refueling. That means we had the aircraft running while the Crew Chief and Gunner rearmed us with the help of some armorers and refueled by field refueling teams. There were aircraft everywhere and it was almost surreal. The A/C took a piss break at this stop, and I would get to get out at the next stop. A piss and a cigarette sounded like a summer vacation right then. This went on for all day and half the night. When we got back to our base we had flown 11.6 hours, an amazingly long day. Our max flight time in a 24-hour period was 8 hours, but you don't leave those poor bastards on the ground because you timed out, so you push on. I wish I could write more of exactly what we did, but I was too exhausted to remember. We had eaten a little of some C-Rations while refueling, but they were hard to take cold and limited water to drink. As I mentioned, these C rations actually were made during World War Two and had been sitting around for a while.

We made it through this day which was encouraging to me and made me realize that no one who had not done these kinds of things will ever really understand. There is the noise of the aircraft and weapons as well as the radios and, of course, there were either three or four of them going. There was the low food and water to have or even have time for. We felt good we lost none of the aircraft we were covering, even though there were aircraft shot down and some shot up. With all the ammunition, I was amazed we had taken no hits. After debriefing we all tried to find something to eat and drink and headed to our bunks. In what seemed like only a couple of hours, we had the same thing again. The only significant difference was we only flew

7.8 hours. It lightened up over the next couple of days, so we were able to bounce back and ready to go again.

HARSH REALITIES

Being new in the war taught me that no understanding of the harshness of war comes until you see it, live it, and fight in it. This fact surprised me in many ways. One of the more challenging situations for me was processing the way the press described the war in contrast to the war I experienced through my father's involvement in it and later through my own.

We saw news reports showing American soldiers, often Vietnam veterans, describing in vivid details their full-blown disgust with the war and how it was being run. At this point in time, the major television networks and newspapers provided the news sought after by most Americans. The exposure provided by most of these media sources was primarily one dimensional, failing to provide viewers and readers with any other way of understanding the war. People soon believed the news reports as the *whole* truth. Then the visceral reactions to the war itself began to feed hatred for the war and by extension for the people fighting it. These reactions came from deep personal feelings ignited by the press reports and the common bond felt by people who saw the US involvement in the war as a complete tragedy. Little in the way of a logical or coherent explanation came from any branches of the military explaining why we were there fighting and dying for people so far away in a land that was that unknown to the average American. Of course, the American people trusted the news reports and some of the news reporters themselves. Many Americans believed they knew and understood the war itself and were righteous in believing that this war was evil. They watched the news depicting friends and neighbors coming home from the war in body bags and atrocities happening to the very villagers and their children our involvement was supposed to be helping. This perspective oversimplified the war.

One problem was that a three-minute anti-war news commentary or selected video of the fighters reduced all the complexities with our soldiers as simply being killers doing the evil work brought about by the U.S. Government. The terribly sad part of this from my perspective was that truly little content was made available for someone to even consider another side of this storm.   Like always attracts like, and once the anti-war messages began to ring true to the American public, that point of view drowned out everything offered by anyone who saw our involvement not only as justified, but also as a moral imperative for anyone believing in freedom. Many of the protesters had never been to Vietnam during this war or at any other time. They were simply regurgitating the words they were fed by so many with no first-hand knowledge of the war.

The US did not depend on an all-volunteer military and used the draft system. Although many people like me enlisted voluntarily, being called to serve was not an option if your birthday happened to fall on a specific date. Some individuals who had no desire, felt no obligation to join, or were morally opposed to the war effort found themselves reporting for duty. Many awakened their sense of patriotism, gathered their strength, buried any disillusionment, and went to war with a newfound sense of purpose.

Others had a different approach when their number was called. They ran away to Canada and hid out during the war. Some did this out of cowardice. Others left on moral grounds when they couldn't meet the formal requirements to be classified as conscientious objectors. Whatever the reason, the defectors received amnesty from President Jimmy Carter in 1977 and were allowed back into the country with no penalty although there could have been serious repercussions for desertion. Only about half of them returned.

In addition, some Vietnam veterans so fervently hated the military and political leaders that were part of the war machine that they began to become extraordinarily strong political enemies of the ruling political parties of the United States. Once again, the number of these totally anti-war and anti-U.S. government war veterans became a strong voice for the anti-Vietnam war side. Once again, the number of these outspoken veterans was truly a very small number of Vietnam Veterans, but they were very loud.

Did the protestors shown every night on the news represent a substantial number of the American population, thereby proving mathematically that the protesters were representative of the American public? The answer would be mathematically no it did not. The number of actual protestors was a small percentage of the American public, but they commanded and received attention. Public support of the war at this time began to diminish although many had supported the early initiatives by the United States under Truman, Eisenhower, and Kennedy, all of which received broad support.

The anti-war movement in 1965 was small, and news of

its activities was buried in the inner pages of the paper, but as Americans came to witness countless hours of television coverage of caskets bearing the bodies of its soldiers and the end of the war not in sight, the anti-war protestors espoused what they saw as new version of patriotism, villainizing those of us who were fighting to free the Vietnamese from the horrors of Communism.

Could we have seen it coming? In 1966 my father was serving in Vietnam, and the news stories were quite different from what he saw there and wrote about in his letters home. When I was serving in Vietnam, the press failed at balancing the reports of the war. They focused on negative things that occurred, providing little news of the successes and for whatever reason, the military spokespersons did little or nothing to counteract the imbalances.

For the first time in American history, the American forces lost no significant battles. American forces did not lose the war in Vietnam as the press portrayed it then and as many historians do today. The last American fighting forces left Vietnam in 1973 with South Vietnam in the hands of the South Vietnamese government. After serious battles raged between North Vietnamese forces and South Vietnamese forces, the South Vietnamese lost to the North Vietnamese. No American ground or air forces were involved or even in the country.

NIGHT MEDEVAC

We had just returned from a busy day and did our standard re-arm, re-fuel, park, and post-flight of the aircraft. We were moving to the revetments for post-flight inspection and preparation for night standby. Before we could even consider eating some food, we were back in the air, escorting a medevac aircraft into an emergency pickup of three severely wounded soldiers from a hot landing zone (LZ) that was close to our base. It was located well up into the first valley about seven miles. The mountains are beautiful, but more so when they're visible in the light of day. It was dark out, very, very, very dark. In 1971, night vision goggles weren't available. Going into a valley that was narrow and having to provide covering fire to the medevac and to the soldiers on the ground proved to be a major tactical issue, and we had no time to work out details before we launched. Of course, communications were going between highly trained and experienced pilots, so I shouldn't imply that we were shooting from the hip, so to speak.

The medevac aircraft commander and crew performed their jobs with excellence. We were communicating with them between our aircraft in the second gunship and the team leader in the first. I had been there long enough now to understand the ropes and work as an effective co-pilot (peter pilot in our lingo) and as an effective weapons officer on the mini guns. Between the external communications, we were being briefed by the aircraft commander (AC) internally. This was my first night gun cover of a medevac flight; moreover, I had flown in the mountains at night only two other times.

The battle was on the edge of the downslope side of the two mountains. In the middle of this deadly fight, with three of their people seriously wounded, the grunts had managed to build a landing zone for the inbound medevac.

The darkness created both plus and minus situations for this mission. We had no moon yet to give us some ambient light. With the hills covered in dense trees, ambient light wouldn't have contributed much anyway. On the plus side, the enemy would have a much more difficult time finding us with any accuracy, a situation that would make it more difficult for them to hit the medevac aircraft. Shooting at a clearly marked medevac aircraft is supposed to be against the Geneva Convention, but who knows what to expect?

We were not going to be able to use our standard types of gun patterns because we had no way to go inbound firing while breaking for an outbound turn. We would have to go inbound and then climb straight ahead at the best rate of climb we could get to clear the hills at the end of the little valley. Then we'd break around above the enemy position and over the top of the low mountains to get in position, hoping to break inbound when the other aircraft made the same maneuver. We shut off all external lights as long as we could see the medevac. Any of the enemies located in the higher elevations seeing our lights would cause us to lose our advantage of the cover of darkness. Of course, once we began firing, the lights from the rockets being fired would show our location. Miniguns shooting about 2400 rounds a minute at three second bursts would be a sight to see, but they probably wouldn't be sticking their heads up looking for us.

We were still half a mile out when we saw the tracers from rounds shot at the medevac. We had all gone to external lights out to make it more difficult for the enemy to see us. This fact increased the difficulty of staying in the right position, increasing the danger; however, we felt we could, and did, do it safely. Maintaining separation from each other and all objects in the mountains at night was hard enough, but to have to do so in the middle of a gunfight was a whole degree more difficult. Add to this issue, once we used the mini-guns and fired the rockets, we lost some of our natural night vision. The medevac started his approach and almost immediately took heavy fire. We had shifted to a modified pattern for coverage for him and our ability to cover ourselves. The lead aircraft rolled hot on the muzzle flashes from the down valley side and had some success because they began firing at our muzzle flashes. This allowed us to narrow down the target area even more. We had decided not to use rockets unless we felt the targets were far enough away from friendly troops. We agreed over the radio that we could hit the back side of the target area with rockets, hopefully catching any re-enforcements rushing into the fight. Our fight raged on taking pressure off the ground forces to load the wounded and catch their breath before jumping right back into the fight. As for fear, I can't say that I remembered much fear because we were too damn busy.

As the medevac departed, he did so with no fire from the enemy. The combination of the ground troops with two gunships in the fight seemed to break the back of the enemy, for a while at least, as they left the battle area in an apparent rush. The medevac made it to the medical facility with no more problems, leaving the wounded in excellent hands. That medevac job was not for the faint of heart. We did not get an update on the soldiers medevac'd out, but we all did our best.

We went back to base, and after a long day, we were able to get the aircraft ready for its next flight. We had time for a couple of beers and to receive mission sheets for the next day for known flights with some of us on standby for the unplanned flights. Another day done in my tour.

This is what we had trained for, and it was the way we had to do our jobs. After several successful night flights, some into enemy fire, we built our confidence to do these tasks when called upon. Any Medevac fight would be done to the absolute best of our abilities. We hung it all out there to get these soldiers to a hospital, a flight that would hopefully save lives and minimize the effects of their wounds. This was our job, and we had volunteered for them, those wounded soldiers who needed our help.

FIRST MORTAR ATTACK

Night was a very daunting challenge. The world flew by at a very rapid pace, and flying helicopters was not an occupation to survive by guessing. If things landed with a guess, that guess had better be based on knowledge, experience, and self-control. Self-control proved to be one vital ingredient necessary to function in this pressure cooker environment. I knew this by listening to soldiers with this knowledge when I was in flight school.

I thought I was ready to give it a try and show the check pilot that I knew the aircraft, that I could fly the aircraft. I wanted to make sure he understood that I knew this was a learning environment. We were expected to perform under fire in a war. Nothing could have prepared us for this. We were in the middle of a war that provided us with "on the job" training. We had to plunge right in. We had done training for inserting and extracting troops under simulated enemy contact, but the key was that the simulation was only simulated contact. Useful, but a far cry from the real deal.

At the time, I could only guess what it would be like when the enemy would arrive. My mind went on and on, playing out various scenarios until I fell into a shallow sleep. Suddenly, I heard the incoming siren, followed very quickly by the first explosion. I saw soldiers running in different directions. Most had M-16's or some other semi-automatic rifle. Many wore T shirts and shorts while carrying 38's or 45's. At first glance, this scene looked like mayhem run wild, but suddenly, I realized that it was organized. Everyone seemed to know what to do and where to go. I saw four of our gunship pilots running at a very rapid pace, carrying survival vests and sidearms, again either 38's or 45's. Within about forty seconds, I heard the helicopter turbines starting to whine and spool up. Within another forty-five seconds or so, I could hear them pick up to a hover. All the while, the mortars kept coming. Fortunately, they seemed to be missing most of the important buildings and aircraft.

The other gun pilots were now heading to their aircraft ready to respond from the cockpit. Shortly after that, I heard M-60's firing along what I figured was the perimeter. To me, this indicated that the enemy had selected their targets, and they were apparently trying to breach the perimeter. The amount of fire from the guard posts intensified with several towers firing. Shortly thereafter, the gunships opened up with mini-guns and then rockets. The noise was incredible and deafening. My heart pumped so hard that I thought it might leap out of my body.

I was trying to figure everything out based solely on what I saw and heard in the moment: an amazing amount of light from the muzzle flashes and some from the explosions. Noise blasted from everywhere.

Reflecting on those moments, I have to say that the training we had must have been surprisingly good. This was my first exposure to a shooting war, and I was in total control of myself. I was captivated by everything going on.

Just as fast as it started, everything slowly ebbed to a stop. All the firing ceased from in the compound. The shooting from the gunships began to slow, and then it stopped all together. I could see the lights on the aircraft as they returned to base and went to refueling and then rearmed in the revetments.

I tagged along with a couple of our pilots as they walked the rest of the way to the flight line. They first asked me if there were any casualties in the compound, and once they were reassured, the mood changed. The pilots told how it went and that they had caught the enemy mortar teams and killed them. No one displayed any pity for them because, after all, they were trying to kill us first. I listened intently as they talked about the experience that had just occurred. I tried to learn all I could from these combat pilots. As everything died down and all headcounts were done and debriefings were completed, we went back to our hooches and talked.

I had a few somewhat uncomfortable moments when the attention turned to me. One of the pilots asked, "How you doing, John? You doing ok? What do you think about what just happened?"

I hadn't really that the time to process it all, but I said, "Honestly, for me this was a lesson on mortar attacks! I need time to think about all this. It happened so fast. Now it's all kind of a blur."

They all seemed to like my reply, congratulating me on keeping my cool and trying to learn. Shortly after that, everyone headed to their bunk to try to get some sleep before the next day would arrive in a couple of hours.

This war would continue on. The things I saw made me proud to be in this company. I knew then that I was going to be serving with people who kept cool under fire. For those who had a role in this, they did their part. The rest went to the bunker to stay alive during all conflict and firing. I saw that the war was real, and I was going to take part in it. Although I needed to rest, sleep would avoid me that night because I had so much to digest.

ACTUAL COMBAT

I have come to learn that certain events that occur in a person's life are seminal events, forever etched in a person's mind. Such is my first day in actual combat. This is true for any soldier, whether they're in the air in contact with the enemy, or on a tanker, whether they're an infantry soldier or in Special Forces. The rollercoaster of emotions, while different for each fighter, is faced by every combatant.

For a pilot, flying into actual combat is an eye-opening event. New sounds come from every direction and must be figured out. New and unfamiliar visuals have to be adapted to. How can one ultimately prepare to be the lead helicopter, flying with the knowledge that you're a target? We couldn't see what was what until the rounds hit something or someone.

My questions were answered by the people I was learning to depend on as I asked them in the moment. I instinctively knew that I had to use discretion when I asked the AC what was happening as we entered the realms of battle.

I asked about the strange noises I could hear, learning that they were rounds zinging by close enough for us to hear them above the din. I asked about the little kicks we felt in the airframe. Those were rounds actually hitting the aircraft. It was unnerving when I could hear the excited voices coming from one of the radios that covered the ground forces.

The brutal noise in battle surprised me, but surely, it should *not* have been a surprise. Had I thought of about prior to entering the battle, I would have known more of what to expect, but it was a detail that had just escaped me until I was in the midst of battle. Noise became a big issue. Add battle noise to helicopter noise. I would have thought that the communication (commo) people knew what to expect, but they had failed to make the facts about noise known to us. Considering they had first-hand familiarity with the level of noise, it might have been helpful had we been made privy to that tidbit of info.

Being in an aircraft gave us more radios to monitor and use. At one point the Very High Frequency (VHF) radio was blaring away, just when we encountered a communication from someone on the ground. It's a generalization, but I saw that if they were older than us and possibly had hearing problems, they tended to speak louder than normal. Radio exchanges on VHF are generally loud. The Ultra High Frequency (UHF) was nearly the same as VHF in sound and effective range. Then there was the FM radio. It did other things, but nothing pertinent to our mission. So, radio noise was almost constant, and the aircraft crews had to monitor these radios continually. In addition to the constant monitoring, one more task for the crews proved to be an absolute necessity and most essential. These encrypting radios could cost more than a few lives should one of the bad guys get their hands on one of them. If we were shot down or set down for in-flight emergencies, we had to destroy that radio.

It became clear that destroying that radio was more

essential than all of us escaping. In some cases the AC kept a hand grenade destined for the sole purpose of destroying the radio. Such an act, if needed, would involve pulling the pin and dropping it into the fuel cell, an act that in most circumstances would destroy the radio *and* the aircraft.

The other good possibility was to have the other gunship destroy it. Serious stuff for a profoundly serious radio. Some of our aircraft had these radios for some specialty missions. Going into combat meant we had mission information, communications frequency, and details regarding when to make contact. We had knowledge of the ground forces and how they usually operated.

Upon approaching the combat area, we established ourselves as the guns working with the ground troops. The AC would quickly brief us as we were getting ready. As one of the two gunships (always two except the few times we had a "heavy" team or three guns), we had listened to what information was set up for the mission, including the patterns we would use and any additional necessary information. Then we were ready.

We joined up with the slicks that were inserting the ground fighting forces or "grunts," and we were inbound and about one minute out before I knew it.

As the co-pilot on this ship, I worked the miniguns. This was a big responsibility because the amount of lead going into the enemy position was devastating. The slicks were on final, and the lead gunship was inbound doing recon by fire, which meant they were putting lead where the enemy might be. As he reached his point to turn, we were inbound with the slicks and covering his underside as he turned, a very vulnerable time.

We were still about fifteen seconds from our point to turn back parallel to the inbound course. The lead gun would be covering us as we turned. The lead slick started taking fire from his three o'clock position. We were on the outside of the slicks. So, we had to adjust to the enemy location inside of our pattern.

In that moment, the lead turned to move inside, and we were going from a left-hand pattern to a right-hand pattern. There was no information given except a quick remark that said we were inside now.

About fifteen seconds out, I found some targets that I could hit with no threat to our own aircraft. I was dead on target, and we dispatched at least four and probably more that were in the tree line. In the sharp turn to the out-bound leg the crew chief and I spotted what looked like a squad of NVA soldiers there. They must have been out of position. They must have believed that we couldn't get them while in the turn, but the gunner opened up on them with his M60 machine gun with deadly accuracy and took out all of that squad. So far, we knew we had taken out eleven enemy soldiers. The number we couldn't see was unknown. The lead gun had gotten about ten, so they were basically confirmed kills. We had caused a great deal of harm, probably long-term harm, to the enemy. Hopefully, this would be a factor in easing any planned assaults on our ground troops.

We continued the fight for the rest of the insertion with very little enemy contact as the last slick went in and dropped off the last of the grunts. We all headed back to the LZ for a debriefing. As we were on short final, a couple mortars landed close to the LZ. The enemy had planned a surprise for us and had they had better accuracy, we would have been in a position that would have been almost impossible to avoid. We couldn't find where the mortars were set, and fuel was an issue if we kept looking, so we were released to go rearm and refuel.

Overall, the whole thing created a situation that we would have liked to describe as the "standard" mission, but as I was finding out, a "standard" mission was an oxymoron. Too many moving parts and personalities existed, so some unwanted actions would surely be part of the operation.

Add to that the fact that the enemy was always trying to disrupt our plans. In this mission, we took out many enemy forces while losing none of our own ground forces or aviation assets. We returned to the staging area and waited for further orders. It appeared the grunts had it under control.

Just then a call came in to see if we could be released for a battle that was starting. This was a ground operation that apparently was growing to such a point they needed guns and slick resources, and they needed them *now*. We left the staging area for this next area which luckily was only about fifteen minutes away.

As we arrived and started our descent to the staging area that they had just set up, we were bounced for a section of this activity. The enemy had stashed some troops in the woods and kept them there out of sight until the grunts walked right into their trap. From the air we could see the enemy starting to surround the US forces. We were cleared at once into this fight, and the fight was on.

We were doing well taking out many of their reserve fighters, which quickly shut down this section of the battle while they tried to regroup. Unfortunately for them, we could see everything that was going on. We were able to stop their plans to reform their force and carry out another attack. We were able to stop them in their tracks.

All of this kept us in the battle until we had to rearm and refuel. In the time since we were first asked, some commander was able to get ammo and fuel blivits brought in by chinooks, a situation which allowed us a quick rearm and refuel and head right back into the fight. Thankfully, one of the planners saw the possibility of needing aviation assets and, therefore, the ammo and fuel had been scheduled on standby. The enemy had reformed while we were off station and apparently had troops in an underground tunnel system, so when they got ready, the fight was on for real.

We spent several hours in this event, and we were able to cripple them with our firing and artillery firing from some miles away. When it was over, we went back to our base to let the crew chiefs look over the aircraft. Surprisingly neither of us took any hits from all the firing. The enemy had done their best, but they did not hit us.

Our day had started out calmly enough but turned into a high-pressure day's work. We went back to our base when released. The day was not over. Logbooks had to be filled out and after-action reports written. The aircraft had to be post-flighted and cleaned. After it was washed, we rearmed it, preflighted it, and set the aircraft up for a combat start. This meant that we used the checklist and stopped when the throttle was set for start. The battery was turned off with the aircraft set up to turn the battery on when the start-fuel system was ready to hit the starter and strap in. This allowed us to get off the ground in only one to two minutes after arriving at the aircraft. We would be ready to go again when the time came.

LEARNING TO DEAL WITH IT

I met up with the Aircraft Commander (AC) heading out of our quarters and to the aircraft. As I followed his orders and preflighted the rotor head and rotor blades, we had little time to talk. He preflighted the weapons and engine. I then went to the tail rotor when the Crew Chief came up to me and introduced himself and the gunner. No time to talk. We strapped in and got everything set up. The AC started the aircraft. The other two gunships were also starting, and in what seemed to be no time at all we were moving into position. These gunships were so loaded that we had to bounce off the skids a couple of times to move into position. The Team lead made the radio calls to get us going. As we departed, other slicks and guns were getting started for their day. It was all done at what seemed to be hyper speed, and my head was spinning as I tried to keep up.

Once we were up and in formation heading towards our Landing Zone, at last we had time to talk. The AC did quick introductions and an apology for having to meet this way, but this is war, and you do what is necessary and adapt. He explained our mission which was to fly gun cover in escorting and extracting some Army Republic

Vietnam (ARVN) troops in a major event of post-Lamson fighting. It had been intense, and overnight the NVA started getting the upper hand. They apparently had many reserve troops to replace any that were taken out.

This was a last-minute mission, and thankfully I had gotten check rides and orientations out of the way yesterday because that gave the platoon a much-needed extra gunship. We had a couple of minutes of Bs'ing before we had to get ready for the battle. We test fired all the weapons shortly after takeoff, and we were good. I remember how I felt through all of this, and surprisingly it was not fear. I was going over in my head what my role was in this, and I was seriously concentrating. As we started our approach to land, we were bounced to a specific spot to cover slicks going in to extract troops. We were a team of three or a "heavy team" and the patterns were worked out.

Then suddenly, we went to work. It was a crazy world of radios blasting and calls being made in much activity since basically all of the guns were going to work early this A.M.

This area was "hot," and we heard many calls made, especially "taking fire." We answered and covered as best we could some very frightened troops and some very intense slick pilots' calls. I have no idea how many rounds were fired at the helicopters, but it was an exceptionally considerable number of small arms and machine-gun fired ammunition.

We made it through that first lift, and suddenly we were in a pattern and working hard. Once these extractions were complete, we had to stop at a field fuel and ammo stop to rearm and refuel. We took a cigarette break on the way, and then stopped for hot rearming and refueling. That means we had the aircraft running while the Crew Chief and Gunner rearmed us with the help of some armorers. The field refueling teams us. Aircraft dotted the landscape everywhere, and it was almost surreal. The A/C took a piss break on this stop, but I wouldn't get out until the next stop. A piss and a cigarette sounded like a summer vacation right then.

The battle continued all day and half the night. When we returned to our base, we had flown 11.6 hours, an amazingly long day. Our maximum flight time in a twenty-four-hour period was eight hours, but you don't leave those poor bastards on the ground because you timed out. So, you push on.

I wish I could write more of exactly what we did, but I was too exhausted to remember. We had eaten a little of some C-Rations while refueling, but they were hard to take cold, and we had limited water to drink. We made it through this day, and I discovered that our survival was encouraging to me. It made me realize that no one who had not done these kinds of things will ever have the capacity to really understand the experience. The constant, unrelenting noise of the aircraft and weapons firing, and three or four of the radios squawking created a palpable cacophony. We felt good that we lost none of the aircraft we were covering even though some aircraft suffered being shot down and some were still flying but shot. With all the ammunition firing around us, I was amazed we had taken no hits. After debriefing we all tried to find something to eat and drink and then headed to our bunks.

In what seemed like only a couple of hours we had the same thing all over again. The only significant difference was that this second time, we only flew 7.8 hours. Things lightened up over the next couple of days, so we were able to bounce back to be ready to go again.

Being new in the war taught me that there is no understanding of the harshness of war until you see it, live it, and fight in it. To a faulty, the American people trusted the news reports and some of the news reporters themselves, which led to so many Americans believing they knew and understood the war itself and were righteous in believing that this war and American's involvement in it was evil. The terribly sad part of this is that it seemed to me that so very few even listened to the other side of this storm.

Each day that I flew I learned more about war and combat and saw how people changed that fought these wars. It wasn't long before I started seeing the change in myself and realized that each incident began to build up and soon, I was not the same as the person I was when I arrived there.

LESSONS LEARNED

1. When I decide I want to do a certain job or work for a specific company, I need to research it as much as possible. Based on available information, I would look at the places this company works in and find out why they are there. Once I'm satisfied and have done my due diligence, I'll see if it looks like they're telling the truth.

2. I need to make sure I can make a positive difference in the company. I need to know how this country operated and when the change of government may have occurred.

3. Once I have done these things, I'll see if I know anyone who works there, hopefully at the same location and in a similar job.

4. When I feel ready to take the job or leave the opportunity, I need to make sure I have learned what others have said

about it and recognize that I have almost no way to know why they feel as they do.

5. When all is said and done, I should listen to my internal voice and possibly act on it.

6. Once I am there (where there is), I should decide if the environment is what I am happy with. I should relax and work hard and do my best. Whatever happens is what happens, and I will adjust to it.

7. In the event I end up in a highly volatile area, I need to do a risk/award analysis. As a pilot, I volunteered to do one of the most dangerous jobs there can be. For the most part, I like my decision. I realize what's at stake in this war, and I have decided the risk is worth it.

8. No matter what happens, I will live with the results of my decision.

INDIVIDUAL FLIGHT RECORD 1971: MAR - JUN

INDIVIDUAL FLIGHT RECORD , J FLIGHT CERTIFICATE - ARMY (PART II)		26. PL 'D COVERED 1971: Mar - Jun	27. SHEET NUMBER 2-1
28. LAST NAME - FIRST NAME - MIDDLE INITIAL MCCARTHY, JOHN M.		29. SERVICE NUMBER SSAN	30. GRADE AND COMPONENT WO1 USAR

SECTION V - FLIGHT HOURS ACCRUED - TOTAL HOURS FLOWN BY MONTH *(For Local Use as Desired)*

JULY	AUGUST	SEPTEMBER	OCTOBER	NOVEMBER	DECEMBER	JANUARY	FEBRUARY	MARCH	APRIL	MAY	JUNE

SECTION VI - RECORD OF FLYING TIME

			FIRST PILOT FLYING TIME										COPILOT FLYING TIME						
						FIXED WING		ROTARY WING						FIXED WING		ROTARY WING			
							NIGHT			NIGHT					NIGHT			NIGHT	
DATE	AIRCRAFT TYPE MODEL SERIES	MISSION SYMBOL	AIRCRAFT COMMANDER	INSTRUCTOR PILOT	FIRST PILOT	WX INST	VFR	WX INST	HOOD	WX INST	VFR	WX INST	HOOD	CO PILOT	WX INST	VFR	X INST	WX INST	VFR	WX INST	CROSS COUNTRY
a	b	c	d	e	f	g	h	i	j	k	l	m	n	o	p	q	r	s	t	u	v
March																					
NO TIME FLOWN DURING THIS PERIOD																					
April																					
30	UH-1H	C			2,3																
May																					
2	UH-1H	C			11,6						1,9										N
3	UH-1H	C			8,8						1,0		1,3								N
5	UH-1C	C			0,8																
6	UH-1C	C			2,5																
7	UH-1C	C			6,5																X
9	UH-1C	C			2,5																X
10	UH-1C	C			2,5																X
14	UH-1C	C			6,1																X
17	UH-1C	C			2,8																X
19	UH-1C	C			2,8																X
22	UH-1C	C			3,5							1,0									H
23	UH-1M	C			4,8						1,0										N
24	UH-1C	C			3,0																X
25	UH-1M	C			2,0																X
27	UH-1C	C			7,5							0,5									W
28	UH-1C	C			5,5						2,0										N
29	UH-1C	C			4,3																X
30	UH-1M	C			4,3																X
31	UH-1C	C			1,8																
					/////LAST ENTRY/////																
31. TOTALS THIS SHEET					86						6	1	1	1							
32. TOTAL BROUGHT FORWARD FROM SHEET NO					9						↑										
33. TOTALS TO DATE					95						7	1	1	1							

DA FORM 759-1 (PART II)

OLD ENOUGH

As time went on, I was doing fewer and fewer flights as I was becoming a senior gun co-pilot. That ended up meaning that I flew more different missions; I was training to become an Aircraft Commander. This was the first time I was to learn there was at least one Aircraft Commander and at least one and probably two co-pilots who were upset that I would become an AC before them. They felt that the AC job could not be done by someone my age. They had explained their feelings in a meeting none of the other gun pilots were in on. Their major complaint was that I was still technically a teenager at nineteen, and I was no better a pilot than them, so it was a mistake to move me up.

In defense of their feelings, I will say it was a serious job for anyone to do. I am not sure that if I was one year older, I would have been better suited to be an AC. In my view, they were simply whining because they weren't ready to be in the position, but they felt they were. We had a new Company

Commander (CO) and a new Platoon Leader (PL). The normal one-year rotation allowed for things to change like that, and I personally never saw a problem with scheduled rotations. Our new CO was on his second tour and was qualified and seemed to be an excellent leader; I would have no problem following his commands. My new PL was on the last six months of his fourth tour in Vietnam, and this would take him into the first six months into his fifth tour. He had done his first tours as a Special Forces, Green Beret soldier. He had spent the last part of the 60's on the ground in Vietnam. He was probably only about 5 foot 8 and maybe 160 pounds and a serious person, but he did have a sense of humor. He was an excellent soldier, but for some unknown reason, I did not like him. I respected him because he had earned that respect.

We got word from some Green Berets that on occasion we would cover them on a hot mission. He was well respected as an enlisted Green Beret and then as a team leader and an officer. The word we got was that no one spoke badly of him. I could not explain why I didn't like him, and there was no reason. He was nothing but fair to all of us, but I felt that he watched me too closely, probably because of my age. I was a bit of a smart-ass, but I did the job, and I was told I did it well.

After this meeting, with the other pilots, he called me aside one evening and told me about the meeting, but he would not tell me who it was, and I felt that was wrong. He thought the issue would cause a big problem in the gun platoon because he was sure I would confront them. Of course, he was correct. Even at the time, I knew he was right. So, I followed his way, and that was to do the best I could do and not give them the satisfaction of letting them know that I knew about their concerns. I found out later that he met with them again, and it seems he came right out and told them it was a chicken-shit way, that they were going to be watched, but he would sign any request for transfer he got from any of them. He did receive one request, and that man transferred. We did not hear from him again, plus we were not told of the reason he transferred. Many years later, I got the whole story. He also found a way for me to fly as the AC but using the PIC (Pilot in Command). I logged Pilot in Command, but the co-pilots logged co-pilot time, when you flew with an AC he logged Aircraft Commander, and they logged Pilot time. This upset co-pilots that flew with me, but they got over it when I was made an AC after I turned twenty, not much later. I had been a co-pilot for four months when I was made AC.

WRITING HOME

Waking up one early morning, (as usual, sleep was a rare event) I found myself at a loss for much to do. Since this was my first down day in a couple of weeks, I decided to catch up on my reading and letter writing. Actually, my Company Commander had brought up the business of letter writing the day before as I walked across the company area enroute to the club. I planned to see if there was anything edible there. He caught me and said in no uncertain terms, "You're not writing home often enough. That, soldier, is a bad thing."

I was perplexed. First, who knows how often I write home? Secondly, why was the CC making it an issue? All I could do was attempt to explain. I said, "Sir, I broke up with my girlfriend from high school when I was getting ready to go into the Army. Then I had only my family to write to, and I didn't really know what to say. I know I can't go into what we're doing in our missions and no detail about specific missions, and even if I could, what would I say? That I had a good couple of days and got quite a few kills?" I asked. (I would find out shortly that he didn't consider my response appropriate, but I went on, "That after the day is done, we go to our little club to eat and sit around BS'ing and drinking."

After a moment he looked at me, and then in a very controlled voice, he explained, "Letter writing is good. It's good for our morale as well as the morale at home. And for the record, as Commander, you need to know that I am very aware of what each and every man under my command does as far as keeping up with the family or friends at home. I also know what mail is being received from home. Is that clear enough?" he said as he looked me over.

I would be doing my part as soon as this conversation was over.

I answered that I understood and would comply and that I was not trying to be disrespectful, but I was sincere in not knowing what to write about.

He paused for a minute and then said directly, "As an officer in the U.S. Army, you should figure it out right now. If you can't do that, I will have you come into my office every day to write. Then you will read to me what you have written, as part of my Military Development."

"Well, Sir," I said a little taken aback, "I feel that you are far too busy for that plan, so I will definitely work it out myself."

He then explained that he asked the censor whether or not I had been writing and staying within the rules of being in a combat zone. Specific kinds of information could and could not be written in the mail. They did not censor inbound mail, but they did censor out bound mail as a safety matter. I did not know that. I thought it was a thing in the movies and that was wrong. I began writing home every other day or so, and I guess that was enough.

NVA ON BICYCLE

Keeping a gun team very close to the Cambodian border meant that our missions were heavier and more frequent. We had only one slick and two gunships there, so we were very self-dependent. The number of aircraft in Vietnam had decreased significantly in the last seven years. Being more self-reliant than not, we understood our situation; in the event of enemy damage or maintenance issues, fewer helicopters would be available to come to our aid. We accepted this reality, and it didn't bother us. It was just how it was.

In retrospect, the issue is more sobering. We flew missions for the MACV station in Ban Me Thout. These missions consisted primarily of a station commander, an American officer, and two squad leaders as they trained and led ARVN troops. The soldiers were in many fights as the NVA buildup in Cambodia was growing. The American Reduction of Forces left the ARVN troops having to pick up the load. Also, the location was within the proximity to the HO CHI Minh trail, the major supply line of the NVA. We knew when we got bounced to help the MACV troops that it was almost always serious fighting, and it seemed to us that the NVA expansion was outgunning the ARVN troops.

At the end of one of these missions, we discovered we had an electrical issue that could cause us to lose firepower to the miniguns. The decision was made to send us back down to our home base in Dong Ba Thin to secure a replacement gunship and return to Ban Me Thout as quickly as we could. The decision to go "single ship," was highly irregular since guns flew in teams of two or occasionally in a heavy formation of three guns. We did not go on a single ship anywhere at any time. This unusual decision was made because we still had a slick there that could be bounced if a firefight escalated more than expected. The mission could be called, and the slick would need gun support.

Any mission could pop up, from ammunition resupply to medevacking injured soldiers to troop withdrawal when the fight ended. Even one gunship was powerful and could help in any firefight. We needed to be able to cover our aircraft in attacks because in outbound breaks, we were remarkably close to the enemy, and the vulnerable, exposed underside of the aircraft needed to be covered. In this instance, a serious threat of an attack on our small little group was possible where we staged, and we would not leave everyone uncovered. We all knew the risks and mounted up to leave as the remaining gunship and slick crew moved to the aircraft to prepare for any eventuality.

We departed and headed down a mountain pass. We knew enemy troops were in the area. We decided to go with the devil we knew vs. the devil we didn't, so we went down the pass a couple of feet off the ground with all the speed we could muster. We were flying about 125 knots or about 143 miles per hour. That close to the ground the speed was apparent. I could see every blade of swirling grass, and small stones scattering. As we cleared the last twist in the mountain road, as the dust cleared, we saw a single soldier on a bicycle riding up the road. One single guy on a bicycle. At this time, the electrical issue locked the mini guns straight ahead, so they were basically useless.

I just figured it was an ARVN because surely no NVA would be so stupid as to ride a bicycle on an open road. The AC stayed down and fast as we closed in rapidly. I was just getting ready to say something when suddenly, the man on the bicycle reached around behind him and pulled out his AK-47. From my vantage point so close to the ground, I was locked in the moment. I could see immediately that he was pointing it up directly at us, specifically towards my chin bubble. I can't account for the actions that occurred next. Everything was in slow motion on the one hand but moving so fast I wasn't sure what was happening.

Then I heard the sound. Drop a watermelon, and note the sound as it hits the ground and explodes. Helicopter skid meets NVA. We were turning in a fast, very low steep turn back towards this soldier. He was lying on the side of the road, his bicycle twisted beside him, and he was missing the biggest part of his head.

The truth of the incident was that this move by this NVA was unexpected and basically stupid. We had not moved towards him nor moved the miniguns at him when he went for the shot directly into the cockpit. The AC had been through one tour and back for his second, which he had completed. Now he was on a six-month extension. He was as experienced as you could get and had been watching closely expecting this individual to try something. His preparation allowed a successful move by the AC that was executed in a mere second or at the most two seconds.

We had no cover ship, and we believed more NVA were

probably in the area, so we went on towards the company area, which was still about twenty-five minutes away. We talked about what happened and decided to just tell the truth when we got back. As we were on the short final to our base, we were sent to join a gun team that we saw departing to make a heavy team for a battle going on and getting larger, apparently. We stopped, refueled, rearmed, and left in very few minutes. We had rockets, and we had mini guns in the stowed position, which meant moving the aircraft to line up on the target instead of moving the guns that way, but they fired.

We joined up just in time to get the plan of attack over the radio and went to work. About three hours after dark the miniguns started moving, so it would be better if they continued working. We flew all night, refueling and rearming several times. The battle turned out good for our troops who were able to rout the enemy troops.

Then we returned home. The skid at the mounting point started weakening during the night and then broke about halfway off. It was right where we hit the NVA soldier, and we had to remove that part of the skid during refueling, which altered the flight characteristics but not dangerously. When we were done, we went to the revetment without refueling and rearming since the aircraft required a replaced skid.

As we exited the aircraft after a tough day and night, the crew chief started cleaning blood and brains off the tail boom. The CO came by to see how we were doing after such a difficult day, and he saw what the crew chief was doing, so we were separated, and statements were taken. We were sent to our bunks to rest while the investigation started.

After about three hours of sleep, the AC and I decided to face the music. As it turns out, one call to the MACV unit sent their people out to search for the body. They found it just where we said it would be. They recovered a pack he was carrying with paperwork translated to inform them that a unit they were unaware of had crossed into Vietnam and was preparing an assault on some town.

We were cleared, and the AC was commended for his quick thinking and reactions to prevent us, me in this case, from being shot through the chin bubble. We were also commended for being able to use this aircraft which was sorely needed then. Some fundamental lessons were learned, but even before that, I still could see the whole event happening again and again. Had the AC hesitated, we would've had a much different outcome, quite probably for me.

In the time I had been in the country after these long months, I had adjusted to the facts of being shot at and even the aircraft taking hits. I knew for sure that except for the grace of God, we could have been hit or even shot down and captured.

These ideas didn't stress me much because we had been taking fire a lot, and we were seeing tracers clearly at night as the enemy tried to kill us. There's no better way to say it, but for what it was. This time I could see the NVA looking at the chin bubble, and I am certain he was looking directly at me from a rapidly closing distance. This situation made it more personal than how I had viewed things so far, and it sent a chill up my spine. It still does, even today.

NIGHTTIME ENTERTAINMENT

We had entertainment at the Officer's Club generally about once a month. Most often, musical groups provided the entertainment. Although most of them were only somewhat entertaining, the longer we were in Viet Nam, the less critical we were of them. The groups sang in English better than I could speak in their Asian languages. Although the majority of the entertainers came from Korea, a few were Vietnamese. I believe now that all of these acts exhibited more courage than we realized or thought about much. I am quite sure that if the NVA or Viet Cong could capture any of these entertainers, they would punish them in the worst possible ways for providing this little light of happiness for the Americans.

Bob Hope came to us on his annual visit. God bless all his people for coming over to entertain the troops. We found out he was going to be close to us several hours before his arrival. I was really looking forward to seeing the show. A break was always satisfying. Seeing Bob Hope in person was big, but something unexpected always seemed to happen, something I didn't count on.

In this case, we had to put two teams of guns covering the show and a perimeter. We kept two teams flying and two teams on rotation when we had to refuel and take a break until their refueling. Of course, I did see Mr. Hope—from about 800 feet above the ground and with no sound. From my vantage point, all I could do was imagine the laughter and the music, the jokes and the warm benefits of this little bit of home coming to us. The wind didn't waft the applause up to me, but with a little imagination, I could hear it and know that everything was ok with me. I felt good to be protecting Bob Hope and his team. After all, they were truly valuable assets of the U.S.A.

On other nights, we were left to our own devices. We generally played cards and drank alcohol, and that's about it. This routine became a way of life that I became used to and, on some level, actually comfortable with.

Of course, we had mortar and rocket attacks to deal with, and sapper incursions became something that became entwined with the million other things that could kill us. In this unit, not too long before I arrived, the sappers had made it through the minefield and wire to blow up a couple of our aircraft. I heard about one incident firsthand.

Our crews ran out in the chaos to protect their aircraft. As one of our crew chiefs ran towards his, he had to make a fast turn around a Conex container 4. As he cleared the corner, he ran right into someone. Bam! This someone was one of the sappers. Both men were armed and as he told me the story, he didn't quite understand the sequence of events as they were happening. He jumped up and the sapper did the same, and they looked at each other right in the eye. They both turned and ran in opposite directions like a slapstick scene in a silent movie.

As he told me the story, he hesitated, looked down at the ground, and shook his head, as if in disbelief. Apparently, he was somewhat embarrassed that he didn't shoot the guy. I asked, "Any negative actions from anybody for the way it turned out?"

He answered, "No. To tell you the truth, I think everyone seemed to understand. They just let it go." I could hear both a hint of anxiety and relief in his voice. Of course, no witnesses meant that no one would prosecute him, and he had come forward after all. He had a record of solid performance during his tour. This one sudden twist of fate that brought him bumping into the enemy and both having the exact same reaction—to turn and run—unnerved him, and continued to do so, more than he had anticipated.

The occasional films provided were a double-edged sword. What may have sounded like a good idea, wasn't one that I embraced. We had an outdoor movie screen, but I never attended to watch one single film, not ever. Call it intuition, but I felt uncomfortable like a sitting duck where a bright light showed the enemy exactly where we were sitting. Fortunately, no one was ever shot there. However, a few of the mortar attacks that happened made me think that the NVA were aiming for the movie goers. Luckily, if they were, they missed.

For me, it was usually beer and cards, but as good as that sounds, it became old. We did have television, though. We had one channel: Armed Forces Vietnam (AFVN) which showed a variety of shows, each one geared to a particular group. We had westerns, soul music channels, and my favorites— war movies and American TV shows. For some reason, I found those options amazing and entertaining. We spent time just sitting around, watching, and critiquing them. Sometimes depression would hit when I least expected it, and I'd start feeling down and trapped. I was always grateful that I had buddies around who always helped me get right again.

When we started flying out of Ban Me Thuot, we had a change of activities. Only our two crews, a couple of tower operators, and a couple of Birddog spotter planes with one pilot and one mechanic lived there. In addition, we had a cook and a commanding officer who appeared to hate helicopter crews. Apparently, he caught hell when he reported a crew for some imagined failings, so we never had to deal with him.

The bright spot in this location was an actual bathroom with running water and toilets. Before, we had nothing even close to that. Here we had an Officers' Club. I use that term lightly because it was only a wooden building about 500 square feet with a bar and a couple worn out stools. We had no refrigerators or ice machines or juke box. In this sad excuse for a real Officers' Club, we were happy to sit there and drink warm or hot beer. I didn't or couldn't express it then, but I enjoyed these moments that made me feel alive, whether the beer was hot or not.

This location was a post you could leave, and it had a couple of bars deemed safe for a group of us if we were armed. I still liked going there because it was close to the Ho Chi Minh trail with a lot of enemy activity. I noticed a direct correlation based on our views of the trail. As more and more American troops were standing down and leaving, increasingly more North Vietnamese troops were training and preparing for their major push to take over South Vietnam as soon as the Americans were gone. That observation on our part was accurate based on what happened in 1975 when South Vietnam fell to the Communist forces. This event occurred a full two years after the last American combat troops left in 1973.

We had some big fights from these troops as long as we were in the area. For some reason, they never tried to take over the little compound we were on, but one group of VC did try attacking us one night. We had no guard towers with guards, so we had to be our own defense. On this night someone heard the VC and alerted us. When we went to the fence surrounding the compound, we could hear the sounds they were making. Shuffling, an occasional undertone of whispering, even a slight cough. They were close, but we could see nothing in the inky darkness of the field beyond the fence. Deliberately and with what we hoped was deadly precision we started shooting at the sounds they were making. They departed as quickly as they'd arrived under the cover of darkness. We figured it was just a small group, and they never ventured an attack on us again during our time there.

It was shortly after this that I did one of the stupidest things I ever did. We discovered that the NVA had cut supply lines coming into the city. They even stopped our truck traffic. The disruption to supply lines was causing serious hardships, so orders often called for us to go after the NVA. When we were running low on food, we didn't worry about it since supply could have flown food to us by airplane. We knew we didn't have to experience the fear of not eating. The locals, however, were in serious shape.

In what I can only call a humanitarian effort, we decided the right thing to do would be to go hunting. We were certainly capable of feeding the locals if we could shoot some animals in the field. Then we could fly the food to the locals in their villages. We had no problem having the flight approved. We geared up and took off.

I was happy to do my part for this good deed, so another gun pilot and I went along with the group. By looking for game from a helicopter, we had expansive views of the landscape. It wasn't long before we saw a large deer running across the field in front of us. The gunner hit it with an M-60. Soon after he shot the deer, we flushed out a pheasant, and he shot it, too. It was then that the crew realized we didn't have gear to sling the deer in. We'd been prepared in so many ways for so many situations, but this wasn't one of them. In a flash of brilliance, we decided one person with a weapon should stay with the game to watch over the deer and bird, so no one would come and steal them. We hadn't seen any villagers around, so we decided this idea would be the best way to ensure the success of our hunting plan to bring food to the locals. Stupidly, I volunteered to stay. I thought to myself, "I'll only have to wait here by myself fifteen or twenty minutes. They'll be quick. They'll grab the gear and be right back."

I left the safety of the gunship, taking with me only a .38. I listened as the ever-present rotors chopping the air became a distance beat. Then it disappeared altogether. There I sat. The deer. The pheasant. My .38 and me. All alone in the middle of a big open field. It was unbelievably quiet. Too quiet. Deadly quiet.

After a few minutes, I saw movement on the other side of the field. At that moment, I recognized how seriously stupid this was, but I could do nothing at this point but wait and pray. I turned on my survival radio and contacted the crew. "I need you to know," I explained, "I have some company on the other side of the field. I am not alone out here."

They took off immediately, as did the other gun pilot. When I could hear the familiar and now comforting sound of the aircraft, I relaxed. I called in where I had seen and heard the enemy soldiers. They had circled the field, and the gunner saw some movement, so he put down a lot of M-60 fire.

When the aircraft landed, the crew jumped out and strapped the gear into the sling. I had protected that deer and pheasant, and now it was going to be hauled into the aircraft. At about that time, the other gunship put down some rockets. The earth beneath my feet shook as the deafening roar made the very air tremble when those rockets hit. That was the first time I had been on the ground when the rockets hit. From my usual vantage point in the sky, I had no idea of their fury.

After this interlude, we delivered the deer and pheasant to

one of the small villages close to the airfield as we had planned. The people were shocked but grateful. Despite the bad choice I made to guard our hunt's rewards alone in a field with only my .38, at least we gave the locals some much-needed food. The PR work proved good

STANDBY IN THE FIELD

American troops were leaving at a fast pace, so this fact left a huge reduction of helicopters. Missions were still ongoing, but they utilized the aviation assets in different ways. One of the most popular was to send an aircraft or teams of aircraft close to where an operation was taking place. This change saved the American taxpayers what must have been a large sum of money because of greatly reduced aircraft flight hours and all the things associated with that.

Sitting in a field so close to combat actions was a good bit nerve racking. We would set out in a field found by ground assets. Often time pathfinders would locate a suitable area for helicopter operations, including room for the aircraft and fuel trucks or fuel bladders, as well as armament resources. I found it to be unsettling to sit in a field with little or no security forces. As with everything else in this war, you adapt.

One day while standing by in a field that served as our staging area, some of the crews suntanned like they were on a beach while others played cards. Others took the waiting as an opportunity to write home or read a book. With no, or little, security forces, we found it best to stay awake and listen to the area sounds. We needed to be prepared to act quickly not only for our own security, but also for the ground troops we were supporting.

For lunch time we had some of those "fresh"— from WW 2—C rations. You can only imagine that Cs were an epicurean delight. One way we heated the food was to have the crew chief put the cans into the exhaust stack and heat them from the hot engine. Despite our best efforts, we could make the "C-rats" only semi-safe for human consumption. That being said, if you were hungry enough even the worst of the WW 2 food had nutrients enough to keep you going.

As inventive as we were out in the field, we found another way to make them a bit more edible. We would use a can that the crackers came in, cut holes along the side near the bottom, then punch it with a bottle opener. Then we would fill up the can with some dirt to make an innovative field stove. Fuel was not a problem. We'd drain four or five tablespoons from the aircraft. For safety, we would then move away from the aircraft and dig a hole that the can would fit in with about two inches of space around it. Then we'd light the dirt and wait for a couple of minutes, placing the main dish on the can of fuel "stove," and cook the food. This did make the "C rats" more edible and almost food-like. The purpose of the hole was twofold: to prevent the spread of fire and to mask the smell.

One of the crewmembers voiced his concern that the fire and/or the smell of food would let the NVA find this area where we were staging. I said to him as I pointed at the two huge helicopters in our area, "What do you think draws the most attention? The noise we make when coming in to land and shutting down? And don't forget the noise of the support vehicles including fuel trucks, maintenance trucks, and armament trucks all moving around. Or the fire from our little field stove?" He didn't answer, but he did look like he was thinking about my question.

I continued, "And do you think these enemy troops like the food that comes in "C rats"? Even if they did find us by the food smells, they find it so repulsive, they'd run." I think he was convinced, but I continued, "If we have a particularly *bad* day, eating this stuff cold will contribute to troop depression." Martha Stewart would've been proud of the way we cooked on those stoves.

Whether or not the mortar round that did hit had anything to do with the stove or not, nobody can say for sure. We were in the staging area heating up our food when a mortar round hit in our area, but luckily, it caused no damage. The fact that it landed was enough to make everyone take immediate action. We all jumped to our feet and grabbed whatever part of our flight suits we had removed, putting them on while getting the aircraft started. During this mad rush, the gunner was collected enough to cover the stove holes with dirt so as not to spread the fire. As all of this was occurring, the mortar fire was continuing and coming perilously close to the aircraft. As soon as we were all onboard, we departed the area rapidly.

Once airborne, we were all searching for the mortar tube. You'd think it might be like finding a needle in a haystack, and it might have been but, luckily, they fired off one more round. They apparently hoped to hit the fuel truck or armament truck. Based on where we first saw the round coming out of the tube, we ascertained the exact location. I was closest to it. Without the need for any discussion, I made a very tight hard right turn, and we fired a burst of minigun that got close, but not close enough. Even with the hard turn, the gunner couldn't make a direct hit. Within seconds, I had the aircraft pointed right at them and fired a two-pair (four) rockets directly at them. We were almost too close, but the rockets had enough room to arm, and they made a hell of a nice explosion of that tube. The mortars must have been right beside the tube. With an explosion this big, when we went right next to where they had been, we found no surviving members of their crew. In fact, we saw no trace of people at all, so the explosion must have vaporized them.

For us, things turned out fine.

TIME GOES BY

I had been in the unit for almost six weeks when I decided I needed to find a nice quiet place to review what the last three months had shown and taught me. I went outside and headed to the ammo storage area. I figured it would be quiet there with no people around. One thing I learned since joining the Army was that I was seldom alone. I enjoy time alone, and I realized I missed being able to just sit and think. In some ways, just sitting and thinking could be a double-edged sword. On one hand, with peace and quiet, I had to keep a handle on thinking things out too much. Trying to decide what sense it all made was a true exercise in futility. Making sense of war is nonsensical.

* * *

I sat down on some sandbags with empty ones as a cushion and got comfortable. I looked back at my arrival and then the move up the coast. I thought about my getting into a gun platoon while bypassing the six-month slick time.

I had nothing but respect for the slick pilots and what they dealt with, but guns were where I wanted to spend my tour. I went over the indoctrination into flying in a war and heard the noises that accompany an active area where we were flying. I could not allow myself to dwell on the missed rounds shot at us. It was hard to understand how difficult it was to shoot down a helicopter. I'd be looking out the window, and often rounds flew by that I could not see, but I knew they were there. I looked back at some insertions we had done while I was in a slick and later found out that we had taken some hits; yet the aircraft stayed together, and so did we.

* * *

I looked back at the night flights that I had done in the mountains where we had virtually no light. When the moon was full, and we were out there, we could see better, but on a moonless night, everything was just stinking black.

I learned little tricks that can make it just a bit safer, but it remained a job that was so dangerous, I didn't spend too much time pondering it. We learned how to do it because we had to. The Viet Cong (Charlie) believed they owned the night, but we were teaching them otherwise. I know I came back from some mind-bending dark flights that went wild when we found the enemy and the fight was on. When we got back to base all I wanted was to sleep. Since I had been in Vietnam, about three months, sleep was a name not an activity. I did not sleep much at all, and soon I was a full-blown insomniac. I tried to hide that from the platoon and company leaders because I didn't want to be sent home because I couldn't sleep. Self-medicating with beer was an acceptable way to get some sleep. I didn't have to get falling drunk for it to help, and I didn't want to get a full-blown alcoholic way of life, so I tried to keep it to something semi-sane.

As a side story here, we had a small liquor store (class six) on our compound that was staffed by a Vietnamese man who had been vetted to be safe on the compound. I went in one day with my ID and ration card while in uniform when this clown checked my age. You had to be twenty-one to buy liquor, and I was not twenty-one. He refused to sell me liquor. It made me so mad, I almost pulled my sidearm and shot him. I could not believe this jackass was refusing to sell me booze. After a shouting match, (I was yelling in English; he was yelling in Vietnamese), I finally left and fetched my platoon leader. He went with me to the store and tried to reason with that little ass, but even he couldn't make any headway. He finally used my ration card to buy a couple of bottles of scotch that I put in our bar, and I paid him back. I made a point of going around there every week or so and just stared that little bastard down, but he never gave in. From then on, I had to get someone to buy booze for me. I rarely drank hard liquor. I was a beer drinker although now I was learning to drink scotch.

And about the beer. We had some Miller and Schlitz and other third round beers that had been on the docks in Vietnam so long that the tops had started to rust. This was before all aluminum cans and pop tops, so we had to use a can opener. We drank it away from our lips because of the rust. I finally understood: WAR is HELL.

That night, I also thought about the food. We didn't eat in the mess hall much because they mortared the mess hall a few times, and with all the tables and chairs you could not get out of the building when the mortars started coming in. We ate what canned food and C- Rations we could get, and with a hotplate someone received from home, we cooked what we could to eat. One of our favorites was corned beef hash and beans and, on occasion, when somebody could get them, mixed with eggs. It is no wonder all the pictures of pilots showed them slender and in good shape. It wasn't the exercise; it was the food.

I also took time to think about something else. I wasn't in country long when we were called to a little village not far from where we were staging. We arrived at the village in about five minutes, and we were in touch with the grunts on the ground. They had stumbled onto a lot of Viet Cong terrorizing this village. The grunts were just a squad and were too far outnumbered to wade into the fight, but the squad leader remembered we were out there to support them. They had arrived a few minutes before us, and the unbelievable slaughter and rapes and murders of these villagers was well under way. When we flew in, we were just at tree top level because we didn't want to take the time to climb, and we didn't know if the VC had air defense weapons. We could see villagers lying dead in the dirt. Some were running, and the VC murderers were catching them. They heard us and turned to fire at us, which gave the villagers seconds to break away. We lit their asses up.

At that point, "thou shall not kill" never occurred to me. I wanted to kill every one of those cowards. Brave men don't wage war on unarmed men woman and children of a small rice-growing village. It took only a few minutes for us to kill a lot of them, but some got away, and the grunts were in hot pursuit. I felt like I could vomit looking at the carnage of villagers the enemy had created, and I wanted to find more of them to kill. I believe some people do not deserve to live, and they should be killed to prevent this kind of a horrible attack. We flew back to the staging area to rearm because we had gone full firing at these little monsters and had put a lot of ammo out.

I now saw what I knew was true. This never made the news, as none of these atrocities I saw first-hand did. This seemed to me to be a manner of aiding and abetting the murdering bastards. It seems hard to understand that when Callie's killers murdered villagers, it was on the news everywhere even here, but this did not even make it to the news in Vietnam on Armed Forces Radio. To my way of thinking, the thundering quiet from the news organizations allowed this to continue.

It was after this that I began to see the change in myself,

and I liked it. I had learned to fight until they were all gone to save the South Vietnamese. When we returned that day, I found they did not make enough beer to wipe these pictures out of my mind. These are events I will always remember with no way to make them go away.

LESSONS LEARNED

1. I must know my enemy and their uniforms to identify them from a distance but at quickly closing speed.

2. I must never become so transfixed by a weird changing event that I forget what I'm doing and what I must do. This Aircraft Commander kept flying the aircraft and watching what he recognized as a rapidly changing event. Throughout the rest of my thirty-one-year flying career, I remembered his seemingly calm composure, but knew he was calculating possibilities, so he was ready to react.

3. I must remember that I'm in a war zone, and nothing is standard, so I will be ready.

4. I will tell the truth even when it seems improbable. Should it be an unusual situation, an investigation will take place, so I will give the facts.

5. As an aviation asset, rarely only one person will be in the aircraft. Aviation works best when team members learn how to work with the other team members. We worked as a team, and when the crew chief exited and broke off the flopping skid part, we all agreed with the AC that this was the right move. Our crew chief and gunners all kept welders' gloves in the back for changing hot M-60 barrels; these gloves gave him the protection to break this skid the rest of the way off without injury.

6. No matter how quickly things change, I learned to keep my composure because that gives me better clarity of events.

7. It is OK for me to look back at events again, so I can learn from them, but I shouldn't dwell on what could have happened. I should, instead, focus on the positive outcome and that I was taught something valuable.

INDIVIDUAL FLIGHT RECORD 1971: MAR - JUN

INDIVIDUAL FLIGHT RECORD Avn. · FLIGHT CERTIFICATE - ARMY
(PART II)
For use of this form see AR 95-64, the proponent agency
is Office of the Assistant Chief of Staff for Force Development

26. PERIOD COVERED: 1971: July   **27. SHEET NUMBER:** 3-1

20. LAST NAME - FIRST NAME - MIDDLE INITIAL: MCCARTHY, JOHN M.

29. SERVICE NUMBER SSAN

30. GRADE AND COMPONENT: WO1 USAR

SECTION V - FLIGHT HOURS ACCRUED - TOTAL HOURS FLOWN BY MONTH (For Local Use as Desired)

JULY	AUGUST	SEPTEMBER	OCTOBER	NOVEMBER	DECEMBER	JANUARY	FEBRUARY	MARCH	APRIL	MAY	JUNE

SECTION VI - RECORD OF FLYING TIME

			FIRST PILOT FLYING TIME											COPILOT FLYING TIME							
						FIXED WING				ROTARY WING					FIXED WING			ROTARY WING			
DATE	AIRCRAFT TYPE MODEL SERIES	MISSION SYMBOL	AIRCRAFT COMMANDER	INSTRUCTOR PILOT	FIRST PILOT	WX INST	NIGHT VFR	NIGHT WX INST	HOOD	WX INST	NIGHT VFR	NIGHT WX INST	HOOD	CO PILOT	WX INST	NIGHT VFR	NIGHT WX INST	WX INST	NIGHT VFR	NIGHT WX INST	CROSS COUNTRY
a	b	c	d	e	f	g	h	i	j	k	l	m	n	o	p	q	r	s	t	u	v
July																					
1	UH-1B	C			1.1																X
6	UH-1M	C			4.5																X
6	UH-1M	C			1.3																
10	UH-1C	C			1.0									1.0							
13	UH-1M	C			4.0																X
15	UH-1B	C			1.0					1.0											
16	UH-1B	C												5.8							X
17	UH-1M	C												0.8					0.8		
19	UH-1M	C												1.8							
21	UH-1M	C			1.0																X
26	UH-1M	C			0.5																
27	UH-1C	C			2.0																
28	UH-1M	C			3.3																X
29	UH-1M	C			1.3																X
29	UH-1M	C			1.3																
31	UH-1M	C			3.5																X
			///// LAST ENTRY /////																		
TOTALS THIS SHEET					26					1		1		8					1		
TOTAL BROUGHT FROM SHEET NO 22					142					7	1	2		31					1		
TOTALS TO DATE					168					8	1	3		39					2		

EVERYDAY LIFE

According to the *American Heritage Dictionary*, "war" is defined as "an organized and often prolonged conflict that is carried out by states and/or non-state actors. It is characterized by extreme violence, social disruption, and economic destruction. War should be understood as an actual, intentional, and widespread armed conflict between political communities. . .."

As I became very much aware, living the Warrior Ethos (see *Preface*) as a way of life applied to our personal and professional lives. For every one of us, it defines who we are and who we aspire to become. It describes how Vietnam helicopter pilots were trained in a belief system that we lived by. Whether we knew it by the name "Army Warrior Ethos" or not, this was the value system that was taught to us beginning in Basic Training. It was ingrained in us in flight school. This training created a group of soldiers with a particular skill, a clear understanding of what we were to do, and the belief system we would work under. It was not necessarily a conscious way of thinking; it was what we had become.

As an Army Aviator in the Vietnam War, we knew we would fly and fight the battle as long as the battle raged. Leaving behind a fallen soldier was never an option. Staying inbound was what you did. In a UH-1H, the mission that involved loading wounded Americans, South Vietnamese Army soldiers (ARVN), Koreans, or other allied troops required staying and waiting on the ground, often as a battle was happening. The slick sat on the ground waiting for the fallen soldier(s) to be loaded. Only the aircraft's Plexiglas and aluminum separated the waiting soldier from the enemy. You could often see the muzzle flashes as the VC and/or North Vietnamese Army shot directly at you. Whirling blades became a major target. There was no leaving, no matter what until the wounded were loaded. It often meant flying through the rounds fired by the enemy. The job required no less than *all* of you, and the aircraft had to get these combat soldiers to a medical facility where the doctors, nurses, and attendants would do everything and more to keep them alive. These slick pilots and crews went inbound multiple times, sat in hot landing zones, (LZ's), and waited until the wounded were loaded. This required a unique personality, and many people are alive today because the slick and MedEvac pilots had that personality.

As a gun pilot, I felt much more comfortable sending rockets and miniguns blazing away inbound and taking the fight to the enemy. Yes, they were firing at us with many different weapons systems, some of which were formidable weapons, but this action forced them to go into a defensive posture, and we took the fight to them. Every round fired allowed the guns to pin down their position exactly and made them as much of a target as we were.

Statistics don't always tell the whole picture, but some provided by the Vietnam Helicopter Pilots Association prove a point. While the enemy shot down and destroyed many aircraft, 3305 UH-1s were destroyed in the Vietnam War with a total number of 5086 helicopters destroyed. Two thousand and two (2,002) helicopter pilots were killed, while two thousand seven hundred and four (2704) non-pilot crewmembers were killed. Approximately forty thousand helicopter pilots served in the Vietnam War. The loss of helicopter pilots killed equaled five percent, or one out of every twenty pilots. Of course, we inflicted many more deaths on their side, but still, Vietnam helicopter pilots were twice as likely to die as any other combat soldier that served in Vietnam.

I personally have a hard time believing the Pentagon's data base figure of 40,000 helicopter pilots serving in Vietnam. The logistics of being able to train that many pilots makes it seem implausible. Even including existing helicopter pilots, it is hard to understand how that many were taught to arrive at that large a number. Of the two thousand or so reported as missing or wounded, how many were unable to return to fight the war?

We did not accept defeat. The ground forces stayed in the fight along with the helicopters, the command structure trying to provide whatever the fighters needed. The attacks also could, and very often did, include fast movers like jets from the Air Force with a large variety of weapons and artillery fire with amazing accuracy. Occasionally, we saw the use of Naval long-range artillery. With various missions all over Viet Nam, when needed and available, we had a vast choice of alternate types of firepower. However, everyone else might have been out working just as we were. The assets that you had were the assets needed to do the job and do it right the first and only time. This was the working environment on a "normal" day. A typical night, after all the required activities were complete, involved setting up for the night flight that could occur. Preparation for night missions and night standby were constant.

The mortars were bad enough, but the rockets sucked big time. There was no sound unless the rocket went past you; the sound was far behind the rocket. In most cases, we heard the siren. Just like in the movies, it would start, and you knew your day was not over. Most of the soldiers on the post ran at all speeds to the bunkers. We had several on the base, but I personally never went into a bunker. We were required to get to the aircraft with all possible speed and determination. We would run to the aircraft because we wanted to be on the attack, but the night standby crew would launch in just a couple of minutes, and the rest of us would stay by the aircraft until it was determined that the standby crew was back inbound to our base.

We aimed to be cranked and off the ground within two to three minutes. That meant when we were assigned to a base with attack helicopters, we knew we would be running through possible shrapnel and larger hunks of metal flying around during the rocket or mortar hit. We accomplished the short takeoff time because once we were done with all the day's activities, we set up the emergency-style startup, so we were ready to take off as quickly as we safely could.

I didn't always agree with what the command determined. One time, a large piece of wood siding was classified as a large piece of shrapnel, but a large piece of siding was taken care of in the after-action report. The first time I had to run to get into the aircraft in this environment, I realized it was just part of it, and we needed to be able to fight if the attack covered a ground event.

I questioned if running to the aircraft was the best choice. Finally, I decided it was better to run to the helicopter, which meant we were not in one spot, so we would be harder to hit. I wondered about my logic, but it allowed me to do what I needed to do. It also made profound sense to be at the aircraft should there be a ground event. There were times when we were not flying much but were still on standby for anything, most likely to cover a medevac. Most gun platoons were organic to one larger unit, which is four-to-two who they flew with or for. An assignment would be made to the 1st Cavalry Division, so we were an Ad-Hoc unit. We were in the first Aviation Brigade, but a separate unit.

We received incoming rockets or mortars a couple of times a month. Now and then, more often than that. We could be called up to help any unit, including those with their aviation assets, a situation that I saw as an advantage. This allowed us to be called more often and use various tactics based on different missions, but we had disadvantages, also. When things went wrong, we didn't have a large group of our own aircraft that could come join the fight to help us out.

The war is busy. In both daylight and night, operations are going on with many night missions for the "grunts," the infantry who did a tough job no matter what weather or enemy being fought. They strapped on their backpacks at sixty to eighty pounds, and they would fight whatever enemy they found.

They were on the ground, and they were there to fight. At times they would need more ammo or food. At times they wanted to receive their mail, which for some soldiers seemed more important than anything else. They had to use what they had until they could be resupplied. They would fight for themselves and their squad mates when things turned to shit out there. They were all out there for the flag, which represents home, and for the American way. Many people have taken that to mean we were trying to turn the South Vietnamese into us, which is wrong. To our way of thinking, the fight was simple; we took to the fight for free and fair elections and the right to allow self-determination for the people of Vietnam.

Everyday life could be summed up with there being no ordinary day. We could have a day down or a day not scheduled to fly. Everything often changes in the blink of an eye. We might be catching some sun and talking with our buddies when an aircraft takes fire or crashes for any reason, and off we go. The teams on standby go first, but then any remaining crews join the fight. If a crew is down or grunts on the ground are in trouble, we were suddenly itching to inflict as much damage on the enemy as is possible.

This war was not about winning or losing ground. It was about body count. As hard as that is to believe, it is true. The South, whom we were there for, was already an established country, but the North wanted to make Vietnam a one-party state, the Communist party. The only way to stop them, we believed, was to cost them so much in loss of lives that they would withdraw from the South and go home. So, for us in the fight, we fought them as hard as we were allowed to. Some of the rules of engagement were strange at best, such as the U.S. side sending gunships out that could not fire on the enemy until they fired seventy-five rounds at us. What we couldn't understand was why we were being sent out as armed attack helicopters with the edict to count every round shot at us. Another rule allowed the North to move food, ammo, and much-needed troops, and to the best of my knowledge, they had no rules of engagement.

These issues were all part of our daily life. We would fly and fight when we were supposed to or stay on the ground on standby or do what we were there to do. Often in the evening, we would sit around primarily drinking beer and talking. Conversations ranged from cars to what type of girls we liked. Sometimes the discussions were more graphic than they should have been. We talked about the war, and we all had a plan to fix it, but unfortunately, the leadership in Washington, D.C., did not care to hear our ideas. We would often play cards for money late into the morning.

One of the things we all wanted was real showers. This is not at all what we had. What we had was a small building, sort of, that could blow over in any windstorm. Our water came from a fifty-five-gallon drum on top of the building. The drum was filled twice a day with the water heated by a small Bunsen burner-type heater, so no one ever got a hot shower. On one end was a plastic sheet wall that concealed a row of wood slats raised up from the floor that had four holes cut with more fifty-five-gallon drums directly under each hole. That was the septic system where feces ended up. You could be in the place using these wonderful facilities when someone would pull the cans out to collect all of the droppings and burn them. Any part of Vietnam that had bases with Americans smelled of diesel jet fuel and crap—nothing but the best.

The gun pilots I flew with never forgot why we were doing this. It takes a particular type of person with the determination to fly into the enemy forces firing at our ground component or trying to take down any aircraft by flying inbound with much of the firing changing to fire at us. The lead was flying into situations that could be very intense. We had a firm belief that these people, the NVA and VC, were people, not just targets, but they would do unspeakable things that were never reported on the U.S. news or to the South Vietnamese people who did not want Communism. As I heard it, most of the indigenous people did not wish a representative type of government either although I don't know how much they understood of our government; the farmers just wanted their paddy field and their family and to be left alone. Our everyday life revolved around protecting these people from a Communist form of government that we understood would make the people indentured servants with a government gone awry.

Everyday life was not to be confused with a typical day back in the "world." Flying days ranged from a few short flights or days of standing by out in the boonies waiting for the call that we were needed. At first, glance, that did not necessarily seem the way to do things when you think how quickly a firefight could begin and go violent right from the get-go. We'd take off with a few minutes flight to the fight with full armament and as much fuel as we could handle, given the temperatures, humidity, and density altitudes, all of which affected aircraft performance. Given a choice, we would take off with maximum ammunition. My aircraft, which was an "M" Mike model, could carry 6,000 rounds of minigun ammo and 3,000 rounds of M-60 ammo (same ammo) for the M-60 in different trays. Add to that fourteen rockets and either ten or seventeen-pound warheads. The extra ammo was due to the larger turbine and some other mods of a "Charlie" (C) model. The extra fuel and ammo were essential to what we all did—go ready to fight and stay as long on station as long as possible. I never went into a fight wishing I had less fuel or ammo. Flying into the mountains at near max gross weight required the pilot to be flying constantly, no relaxing. Into the fight, all of the crew were very busy; the pilot flying had his hands full flying, plus watching and listening to what was happening, directing the attack by the guns and coordinating with the ground units as to where the guns needed to attack. The co-pilot was busy directing mini-gun fire exactly where needed. These guns put out so much lead we had to be on target, or we could wreak havoc on the ground troops. The crew chief and gunner were busy protecting the aircraft with their M-60 firepower, as well as going on target with the mini-guns if needed. The gun teams were like a well-oiled machine because mistakes were deadly for our teams. I honestly believe that deep down inside, all gun pilots dreaded the thought of shooting friendlies.

One night we had gone to the club for some semblance of food and to discuss the day's activities. A few minor issues required the slick Aircraft Commanders (AC) and Gun ACs to work things out. There had been no harm to equipment or injured personnel, but it could have swung that way. When the discussion ended, we had no hard feelings, and we all learned the danger of becoming too comfortable when all was going well.

That was "fixing to" change in the near future. We had two new pilots who were at the start of their second tours. Both had flown slicks and then guns; they had even been in the same company and had flown together, but that had been about eighteen months earlier, and things had changed. They were both rusty but getting it back together quickly. Somehow, they were both on night standby as ACs. It should not have happened, but it did. No one caught it until the guns were bounced to locate and kill some NVA. It led to a very intensive firefight. It would come out later that the NVA had set a trap up for the gunships they knew would be out working to locate the mortar site. A very intense night firefight ensued, and the gunships won that fight. It was a hard fight. We could see the tracers from both sides, far more from the guns than the NVA. It was a very costly night for the enemy. It was interesting to follow it both visually as well as on the radios. Operations set up a speaker for us to listen to. These newly arrived experienced Captains showed they deserved the Aircraft Commander slots they were assigned to.

My advancement to AC was set back a little but for a damn good reason. They appointed me as the Pilot in Command (PIC). It was fine with me. I had the same duties and responsibilities as the Aircraft Commandeers. The co-pilots were unhappy because they logged copilot time instead of PIC time since only one pilot could be the PIC. As the saying goes, they could get over it or outgrow it, but any problems would be dealt with seriously. The Army is not run as a democracy. There were obvious leadership duties and responsibilities. The platoon leader that moved a pilot to a PIC or AC position was solely in his realm of responsibilities, and when someone tried to change it to a popularity contest, they were usurping the role of the person in charge, and bypassing an immediate supervisor was dealt with very severely.

The two renewed gunship Aircraft Commanders received the call and took off while we all were running to the aircraft from dinner, and they were ahead of us. We all stopped at the Tactical Operations Center (TOC) since our aircraft was no more than about seventy-five feet away. We sent the co-pilots to the aircraft to get ready to start. At the same time, we listened to what was happening. On the second inbound run, the scream went out, "Cease firing, cease firing, you are shooting friendlies. We all froze in place and listened for what was next, hoping it was an error. The calls went out for the guns to get out of there, return to base, and stand by.

This exchange caused a great deal of activity by the bad guys, who heard the absolute quiet caused by the screaming and silence of guns not firing. Somehow no new deaths occurred after the firing stopped, and the grunts took their rage out on the bad guys and drove them off. As the guns landed, they were met by the Commanding Officer, The Executive Officer, and the gun platoon leader. We were kept away from the flight crews who were taken to the mess hall for all of the debriefings. In a short time, the Battalion Commander was flown in with some extra Military Police. It was unbelievable how quickly the investigations were starting. Our company was shut down pending investigations that seemed to be looking for a guilty verdict on some outlandish charges that I won't even say I remember all.

The next day it was all over Vietnam that we had shot friendlies. The decision was based on how the pilots explained what they saw and insisted they did not shoot at the friendly locations. One intelligent senior investigator got smart and ordered autopsies of the bodies of the friendlies that were killed by the so-called friendly fire killings. However, I believe they were to be done back in the States in San Antonio. We were shut down, and the whole company went through an unannounced inspection that went to the bottom of everything we did. The company pretty much aced it, so some of the heat wore off. Three days later the results were in. The metal in their bodies was not the same metal used in American rockets or mini-guns. The pilots had it right. They fired rockets where they had seen gun flashes; at the same time, the bad guys fired what were believed to be B-40s at the grunts.

In the middle of a firefight, everything is so hectic that things happen so quickly it is hard to be sure of precisely what may have occurred in a deadly event like this. Thankfully, cooler heads prevailed.

I expected the monsoon season to be all about the rain, but luckily, it held off until this ordeal was over. I had never seen rain like this. It was so hard for days that we were stuck in our rooms but better than being out in the jungle as so many infantry troops were. We had a break in the rain after a long time, and two crews of us were to go up to Ban Me Thuit for gun cover in the event of NVA activity in this mess.

There were no real problems until my team got caught in a severe rainstorm in a mountain pass. The rain was coming down in these very intense shafts. Big torrents of rain in big gray, relentless columns that we could only fly around because nothing flying could have flown through them. The situation was even more dicey because we were in a narrow mountain valley's passage. Looping around these shafts of rain in this surreal feat of nature was unlike anything I had experienced in my short flying career. The experience was truly surreally sublime; beautiful to be there and navigate it but on the edge of real danger. Once back at the BMT site, we settled in for more intense rainfall. We were not disappointed; it came down in torrents and for days. Food became a little short, but we had C rations we could and did eat.

Officers drew money as separate rations so when we ate food from the Army, such as in a mess hall or as was more standard C Rations, we had to pay. The C rations we had were dated back to the 40s. So, we settled for green eggs, ham, and other culinary delights, but it was food. So, we sat in our half-walled half-screened rooms, eating C's with no entertainment. War is, in fact, hell.

SABOTAGE

An enormous amount of ink has been spilled about drug use and the American military fighting force. Mass market movies, popular novels, and even war memoirs have contributed to the misconception that drug abuse was rampant. As a sceptic of most of the media-perpetuated myths related to Viet Nam, I'm not sure how much of the information these stories are based on was substantiated or proven to be true; however, I do have some first-hand experience in my role as the Alcohol and Abuse Officer in our unit. So, the story I tell comes from authority. I may be perceived as generalizing about all drug users, but I base my observations on what I saw happen more than once. Heavy drug users in this war could be vindictive and mean. They could be poised to attack anyone they thought might be "snitching." They weren't above sabotaging our aircraft if the crash could take out a perceived enemy.

In aviation units, many officers have a "Military

Occupational Specialty" (MOS) related to flying. My MOS was 100EO Attack Helicopter Pilot. Due to the heavy number of pilots in our unit, officers had an extra duty. In normally structured units without a large number of officers, these positions would have been filled by a few officers as their sole duty. These duties ran the full range of positions that officers had to fill such as Motor Pool OIC (officer in charge), Maintenance OIC, etc. As I mentioned, I was made the Alcohol and Abuse Officer. I understood the idea that because of my age, I would be able to relate to the younger enlisted men; I do not believe that was the case. Growing up in the sixties, (remember sex, drugs, and rock and roll), I saw what drugs were doing to our younger generation, and I was strongly against illegal drug use.

In Vietnam, it was easy to recognize soldiers who were using indiscriminately. This wasn't true in the case of the smarter soldiers who used at night when they were in their rooms. At times, other soldiers would turn in users who were unable to control their drug use when close to duty hours. I had no problem busting users when we received a tip or even just during a normal inspection to find drugs. (Alcohol was another story. Most soldiers would not be able to drink while doing their jobs. The smells and packaging of the alcohol was a dead giveaway.)

When we caught a drug user, if no one was injured, no equipment was destroyed, or no mission was compromised, we could offer a deal. If they admitted their drug use and explained how the obtained the drugs, they could go to rehab (in Vietnam), and once clean and released, if they stayed clean for one year, their records would be expunged of any information about their drug use. However, if they did not choose rehab, they would be remanded into custody and then court-martialed, or maybe receive an Article 15 and not be formally charged. The very serious court-martial could lead to much more severe penalties than an Article 15. These included jail time, loss of rank, and perhaps loss of pay. Of course, rehab was the smart choice to become clean and stay clean, but not all soldiers in this situation took the opportunity for rehab. This amazed me, but as one soldier, who had been to rehab himself, explained to me, many of the soldiers had been using drugs for a long time and in most cases, prior to military service. They didn't believe they could stay clean and sober and preferred the idea of being released from the military, resuming their old way of life, which might very well include drug use.

No matter what the personal case, illegal drug use was even more dangerous in a war setting. In more than a few cases, I witnessed users become paranoid and attack people they thought would turn them in. As disturbing a thought as this is, some would sabotage the aircraft to get at one of the crew members, endangering all the people on or near the aircraft.

To the best of my knowledge this was rare, but it almost got me one day when I had a mission, and my aircraft was in maintenance for a scheduled inspection. I was assigned to an aircraft previously assigned to one of our other gun pilots. I was preflighting the aircraft, making my way to preflight the rotor system. I grabbed an engine cowling handhold to pull myself up to stand on the hard mounts of the weapons system. Suddenly, the cowling pulled from the aircraft, causing me to fall. The cowling fell on top of me. This may have been human error that happened during the aircraft inspection when someone set the cowling in place without realizing they had done it incorrectly. I immediately wrote the incident up in the logbook and headed to operations to see what aircraft I would be preflighting now. My first thought was a sapper may have entered the compound the previous night and messed with the aircraft, but I immediately dropped that idea. I had no absolute proof, but because of my previous experiences, I figured someone was after one of the crew members and sabotaged the aircraft. As luck would have it, my aircraft had just come back up, and all it needed was a quick test flight, so fortunately, I flew it for this mission.

The mission went on and all went well. When I returned, I was questioned by a board of officers about my observations and report. The meetings didn't last long, and all I had to do was answer their questions, and then I was done. Later, I found out some of the details of the investigation. The perpetrator had removed the cowling and all the transmission cover bolts. He stuffed a rag into the transmission through the small space he could open on it. Then he had put the bolts back in. I guess he thought that rag would get caught in the main gears and destroy the aircraft during startup. About a week later, I learned that they had caught the guilty individual, and he was going back to the states for a trial on serious charges.

Looking back at that incident throughout my thirty-one-year helicopter aviation career, I've often thought but that for the grace of God and the incompetence of the perpetrator, we could have been severely injured or even killed had this caused a transmission seizure during startup or transmission failure once we were in flight. These incidents were rare, but they did happen. In this case, it happened to me.

Throughout my tour, I often wondered how the drugs made it into the compound since we were a closed post. Even as the Drug and Alcohol Abuse Officer, I never found out how the drugs came in. Given the situation, I always remained aware that someone who knew what they were doing could sabotage the aircraft, so I made damn sure to preflight it well.

It was difficult for me to understand how a person could justify sabotage. It wasn't only the enemy out there, but also someone who perceived being wronged trying to kill you. Drug use was dangerous to all of us. The compound was designed to keep people out, not in. Since users could potentially harm American soldiers, we learned to watch inside the compound to keep from becoming victims of users while watching outside the compound to keep it safe from the other enemy. Fortunately, our unit had far fewer drug related problems than many other units I had heard about. Most of the soldiers, at least in Aviation units, were against drug abuse, and every time the dangerous actions of the few came up the other soldiers helped stop them.

HOME BASE: DONG BA THIN / BAN ME THUOT

Our home base was well designed and suited us perfectly. That was an advantage of coming in later in the war. The soldiers before us built the base themselves. The original aviation troops who arrived before us started proving that helicopters brought a whole new dimension to the war. They were the ones that physically built our base. This allowed us to get some rest, so we could stay sharp for the next mission. We had electricity although it wasn't strong enough to run air conditioners to cool down the oppressive heat. We did have television, but we only saw a broadcast called Armed Forces Vietnam (AFVN), which had a station that was non-political with a variety of types of entertainment. The most watched show on our base was *Combat*. We could follow the news source, which didn't have much spin on it.

In some cases, however, virtually no time occurred before we saw the difference in what was reported and what had actually happened. I was appalled by what the press had done to the truth. Once I had participated in an event such as a battle that involved the actions of the American military, I would feel almost physically ill by the convoluted story that was the news report. I discovered a process by which the story released was often twisted and difficult to follow, with the ultimate takeaway being the easiest interpretation of the story and creating something with little resemblance of the reality of the event lodged in our experiences.

Americans had absolute dominance in the air. Being shot down by anti-aircraft fire, as seen in World War II was rare and was mostly a danger to the fast movers and bombers flying up to and into North Vietnam airspace. The amount of ammo expended by ground troops with their AK-47's and 51 caliber machine guns proved to be a real threat to the helicopters. The number of helicopters destroyed by the ground fire was significant and the flight crew that were killed, captured, or wounded was also significant. The best way that I saw to show what it was like was to see the air to ground and ground to air firing at night. The tracers fired by the American gunship mini guns was a solid stream of red tracers. Four non-tracer rounds occurred between tracer rounds, so while the gun fired fast in three-second bursts, it looked like a solid red line to the ground, and it showed green short lines as the enemy fired at us. So, they were seeing only a small percentage of the rounds coming out of the barrels. Of course, we had a huge threat from the enemy, but these Huey gunships were more lethal than they looked.

On our base, revetments that created the retaining walls were made of solid steel planks laid horizontally with sand and then another set of solid steel planks. This system protected the aircraft from mortar and rocket attacks. This same type of sand barriers protected our housing, "hooches," where we lived, but it provided no protection from direct hits by rockets or mortars. There were mortar or rocket attacks every month

I was there, often more than once or twice a month. Standard sirens signaled that we were being attacked.

Being the gun platoon, we were ready, as ready as you can be, for these attacks. The night standby pilots were set to go in just moments, all the way down to sleeping in their flight suit pants with boots on and laced with their shirts and sidearm right next to their bunk. Their aircraft were set to start and spring-loaded to the go position. Once the attack started, they were gone with the idea of being in the air in minimal time, about two to two and a half minutes. A well-trained crew could do this with no problem. The remaining crews would get ready to go and some would head to their aircraft in case more than one team was needed. Those crews that had been relaxing with a few beers or more generally stayed in their rooms. We never went to the bunker, a choice we all made with no pressure from anyone. I know we generally believed we were safe where we slept, and the commanders let us be. Deep inside, I think they liked the fact that we would not let anything force us into a bunker. We were ready for the fight at any time if we had not been drinking. Even then we refused to go underground.

The wild card was "sappers." They were part of the elite Communist units initiating surprise attacks. These guys could hide in the mine field in our perimeter and could somehow stay hidden even through the day. Their goal was to get in under darkness and sabotage the aircraft. A story I was told, which I verified with the crew chief directly involved, made me aware of the imminent dangers sappers could cause. Just before I arrived, a sapper made it through the mines and guard towers to set charges on several of our aircraft. The sappers were running through our camp when one ran right into one of our crew chiefs. They smacked right into each other, the sapper with a pistol and the crew chief with an M-16. Both completely surprised, they looked at each other, then got their wits about them, and both turned and went their way with no shots fired. The crew chief was a little embarrassed, and I didn't know if I could do any better in the same situation. The story did make us aware to be set up to access our sidearms quickly should the need arise.

I grew quite fond of our base and hated to leave it when we stood down. The club we had was not bad and was air-conditioned, so it was a good place to hang out, and they had cold beer. My main complaint was they did not have Budweiser, and what we did have was awful. The cans were rusting for sitting on the docks so long. This was before pop tops and aluminum cans, so I figured, "What the hell— it's a war, and we all have to sacrifice."

The club worked a deal with some Air Force and "swabbies" (Navy personnel) where they swapped something, I don't know what, for steaks. We had steaks so often we actually got tired of them, but the price was right, and they were grilled by a real cook who knew what he was doing.

We did not go to the mess hall to eat often because when we got mortared, the enemy seemed to target it, so it was damaged quite often. Although one morning we had a couple crews there while we were put on standby for a probable attack in Cam Ranh Bay airport, a large American Air Force base. As we waited, we had coffee and talked about different avenues that the VC might take to hit the mess hall. Suddenly, we all heard a missile go by overhead, and then it hit something. We later heard it was the ammo dump. We all pushed at the table to flip over backwards and get out the door. Just as we hit the floor all the overhead lights came crashing down on our tables. We all emerged without a scratch and made it to the aircraft, taking off into the air very quickly. Fortunately, we caught two of the enemy fire teams just as they were escaping, and we got them all. We would catch a few stragglers as they were heading out, so the attack was an expensive one in both manpower and in dollars and delays for the Air Force. We were successful with no damage; we also had a stronger case of "I won't go in there again!"

MILITARY INTELLIGENCE

M any people, including myself, make jokes about military intelligence—or the lack thereof. The intelligence world works with many moving parts, everything from aerial photography to basic observations, to human knowledge and observation as part of the spy world. Not all these resources or the people involved prove to be spies as their designated profession, but they see an opportunity to hurt the people that control the observer's life.

The U.S. Military operations in Viet Nam have had combat insertions in specific areas based on human intelligence. In one case our military establishment had a spy in the planning section of the NVA's headquarters. For whatever reason, I believe he led them to fight and capture some enemy paperwork that was very valuable. To the military establishment's way of thinking, this made him a reliable source of information, so his info did not require a backup of proof. This situation led to an insertion of quite a few troops from various companies' use of assets for a large operation.

We were involved and were briefed on how the operation was going to work. The whole plan was to have a large force inserting into a different area than what it appeared. They believed there was a spy within their operation, so they drew up an insertion plan and set it into action. The reality was that on the morning the operation was getting ready to depart, they had different landing zones and objectives. All of this was known to very few people and one of them was this spy, so he could be surprised in appearance when the plan unfolded wiping out the NVA in that area. Even during the briefing, we, our crews, looked on in disbelief. It amazed us that this was an exceptionally large operation that may go south if this traitor to his country was considered a believable and reliable asset. We were sent out to prep the fake landing areas as if this was where the attack would take place. We were livid that we were used only as part of the ruse.

Long story short, the first aircraft that was doing the real insertion was shot all to hell by enemy troops that were hidden on the approach path to the first landing zone (LZ). All of the Korean forces in that UH-1 were killed as well as three of the four members of that crew. The surviving pilot was almost dead from all of the hits he took but was somehow able to fly the mess of an aircraft back to where he took off.

The aircraft was parked behind us and a few feet up a slope. When we arrived, we parked where we had been and saw that the aircraft was obviously shot all to hell. When we were shut down and post flighted, we walked to it for a closer look. We weren't the only ones; many soldiers arrived to look at this aircraft. We saw what initially looked like transmission fluid draining out, but then we realized that it was blood from the dead and the near-dead aircraft commander. So much blood.

All nine of the Korean soldiers were dead in the back, as well as the crew chief and gunner. The front of the aircraft was riddled with bullet holes. Under no circumstances should that aircraft have been able to fly. No human, no matter their pilot skill, should have survived that amount of lead and been able to fly that aircraft with any precision, much less having the capacity to fly the aircraft back where it started from. We knew the pilot was not going to make it based on the injuries some of the guys saw when they took him out of the aircraft.

Some things you never forget. As we suspected, it turned out their spy was rewarded by the NVA for having set us like that.

ONE FINE DAY OF WATER SKIING

We had been flying day after day without much time off when out of the blue, we were given a couple of days off. In normal circumstances, time off would be welcomed, but although we were big on anticipation, the truth was, we discovered that we preferred action. We'd be crawling up the walls by noon, searching for ways to fill a day of newfound freedom.

In this case we all had time to do what paperwork we needed to do for both our primary job and our extra duty. The three of us were done by noon when one of the slick pilots came by and commented that we were looking bored. We offered various versions of a resounding, "Yes." He was full of information, and with a glint in his eye, he stood up straight and announced, "You guys think you're really smart, but I'm betting you don't know what's going down at the Air Force Base (Cam Rahn Bay)."

We scratched our heads, and admitted that indeed, we were

completely in the dark. He took immense pride in announcing, "Well, guys, they have a special service building where you can get a BOAT with a motor and skis and ski gear."

It didn't take us more than five minutes before we had scrambled around using the field phone to call the Air Force. We told them who we were and what we wanted. A friendly voice on the other end of the line squawked, "Come over whenever you're ready!"

We set up a ride with the motor pool, received permission to do this trip, and off we went. When we arrived, we were amazed to see that no one else was skiing. Not one other person was taking advantage of this incredible opportunity. We did see some Air Force personnel lying on the beach. A few armed guards walked around patrolling the area. We found a specialist, and he hooked us up with the gear. He kept chuckling under his breath, so I asked him, "What's so funny that you're giggling as you're doing your business?"

He replied, "Sir, you all seem to be pretty happy about this, and I am glad to see it. Usually, we get the Air Force personnel on the water, and they seem bored with it. I seriously doubt you'll find this boring."

I thanked him and after a few minutes of instruction off we went. All of us could ski, but we weren't in the league of competition skiers either. We took turns driving the boat and skiing. One guy on the water, one guy driving the boat, and the other watching the skier. We all fell and crashed numerous times, but we had a great time, and for this little while, we felt like normal guys just out for some fun on the water. The setting of the beach and the chance to boat and ski were welcome changes. Something about flying across the water behind a speeding boat charged our batteries and made us feel even more invincible than when we arrived.

At the end of the day, our ride showed up, and we got back into our uniform and military frame of mind with the addition of our weapons. We went back to our compound, which was just about 20 minutes away. We talked to all our guys about our fine time and that the Air Force let us their boat and ski equipment. Then we all racked out because we were tired. Before we left for our quarters, we were informed that tomorrow morning, we had a special type of flight. We knew what that meant. It would surely prove to be very busy and a tough fight.

The next morning the orderly came and woke us up to get ready for our flight. He was direct as soon as we opened our eyes and rubbed the sleep out, "Ok, get up now. And you're going to need your bayonets for your M-16s."

That got our attention. We all got our gear, went to operations, and were handed our mission sheet. I was reading all the pertinent information, when I was stopped in my tracks as my eyes hit one very important word on the page: "Sharks." What? Did it really say "sharks"? Granted, I hadn't been awake very long, but it was taking some time for this word to compute. Then I gave the mission sheet a closer look. OK.

We were on our way to the beach at the air base. To the Air Force Base (Cam Rahn Bay), and we were going there to shoot SHARKS. The other aircraft commander looked at me, and we looked back at the mission sheets. Many large sharks had been swimming in the water, apparently looking for food. Water in this area was shark infested, and the sharks had been there to feed for days. From the Air Force perspective, three guys had gone water skiing amongst them and came out of the water unscathed, much to everyone's surprise. Now the Air Force wanted us to kill the sharks or at least run them off. Suddenly, we realized what the specialist had been chuckling about when we gathered the equipment and set out on these waters in the little ski boat. He either thought we were extremely brave or extremely stupid to ski amongst them. I think I know which.

We took off, flying over the beach and the water. At low

altitude, we could see the gray shadows moving gracefully under the water. Sharks. Lots and lots of sharks. We gave it our best shot, but as it turned out, you can almost never kill anything in the water, and sharks are no exception. The water absorbs the shock and speed of the bullet and the rockets. So, we spent a full load shooting all that ammo into the water, hoping to annihilate the shark population, but it was not to be. We failed to kill even one single shark. Not one floated to the surface with a bullet wound.

We have no idea why the sharks did not bother us the day that we swum and skied among them, but we had a theory. We chalked it up to the fact that on the day we skied with them, they were just not hungry. We did make it back over to the beach on a training flight of our own design one other day. We were in uniform, flying extremely low around the beach, so the specialist who rented us the boat could see the aircraft, and he would know, without a doubt, who we were.

When we landed and approached him, we told him we were coming back very soon, and that we were arranging for him to go on a flight with us to "show him our gratitude." We thought he was going to cry. We left with him believing we were working on arranging the flight. Of course, we never saw him again or bothered him. After all, we were the ones who wanted to ski with the sharks, and who knew whether or not they'd be hungry the next time around.

TALK ABOUT THE WAR IN VIETNAM

Any of us who served in Vietnam have a story to tell. Some might not want to try to write down these experiences because of a fear of rejection or some other real or imagined fear. Many of us who served will talk, but usually more comfortably to other Vietnam Vets because we don't want our families to have to live with learning how seriously tough the experiences were. It seems some former soldiers think they might be seen as weak or as a failure in all phases of life should they share their stories.

I've researched many articles about PTSD or being "tagged" with PTSD. The prognosis covers everything from a risk to your community or family to your inability to hold a job. PTSD is a sign of a mental health issue, and anyone who has it could blow at the least possible event, a situation described in popular terms as "going postal." Cases exist were the illness caused a person to kill or maim everyone nearby.

Some actions might work as a trigger and set someone

off into a furious and dangerous fit. Of course, PTSD not only affects veterans of wars, but also anyone. The right trigger can set any person plagued with PTSD into a violent rage, causing them to cross that invisible line and begin hurting innocent people.

When I returned from Viet Nam, I was sometimes told that nobody cared about the conditions of returning veterans. In reality, the veterans who desperately needed help could sometimes find, and did find, the right people or program. However, it was so much easier not to talk about war experiences with anyone than fend off their attitudes about the war. In reality, for some, a severe shortage existed of caregivers with whom the combat vet felt comfortable.

Talking about the war was not something combat veterans from the Vietnam War wanted to talk about. They often found little interest from non-combatants. This war was different from other wars. We had fought for the body count because it was always the deciding factor in the effectiveness of the operation flown. Body count was what decided if you won or not. This was not what the fighting soldiers believed in (in many cases). We fought for land but would often abandon it shortly after winning it. The tangible victory of taking land and holding it was lost in this war. *How* we were doing was not a factor in the Vietnam War.

Since we don't have a real jungle here in America, it was impossible to describe adequately how the fighting would be. The fighters who had been there awhile, counting down the time remaining on their tour, truly learned on the job. Nobody provided a realistic way to explain what living in the jungle would be like. The "grunts" (not a disparaging name) who fought on the ground learned as they went. They followed orders. The problem seemed to be that someone decided that holding land was a soldier-intensive formula, so we didn't want to go into the fight that way. It escapes me that no one figured out that we could run the enemy out of South Vietnam if we squeezed them out, holding onto the land we won. This would've forced them to retreat since they had no place to hide.

A triple canopy covered everything below the top canopy, or layer, of treetops that concealed the ground from above, so we had no way to envision how to fight from the air to the ground. It was a new world. The steep learning curve and training associated with the aircraft needed to prosecute the war occurred in a combat zone, so spending more time flying around training was not as advised as you might think. Almost all of the young pilots who went to Vietnam had completed just ten or a few more months in helicopter pilot training. The pilots were good at flying and understood their jobs there. So many of the flight instructors at the Aviation Center were Vietnam veterans who could impart the scheduled training and add knowledge about how to fly in the Vietnam War.

After the experiences I had and seeing first-hand the Vietnamese people's struggle to maintain a free way of life, it became harder and harder talking about the war. People had been against the war, and many had actively protested the war. I personally went through situations where I encountered post-Vietnam disgust. I returned from service to be called many disgusting names by people who were, in fact, ignorant of the truths of the Vietnam War. I learned first-hand that public anti-war sentiment was fed and stoked by journalists. Since I served and fought in that war, I saw the true outcomes and benefits of our presence there. I found it hard to tolerate the people that I considered as truly misinformed.

At the time, the most trusted man in America was Walter Cronkite on the nightly network news. He and his counterparts told the Americans a version of the truth, even when it was *not* the truth as I perceived it from my experiences there. Many individuals armed themselves with only what they heard on one political facet of the news, primarily with an anti-war left-leaning agenda. Many people had no interest in listening to the other side. They transferred their hatred of war to the soldiers who had fought, sacrificed, and died in Viet Nam for what they believed to be the noble cause of freedom. Anti-war demonstrators had not been there to see what I had seen with my own eyes. They had not seen what the NVA and VC had done to people who attempted to find the strength to rebel against the Communists' overthrow of their democratic government. Freedom had been their way of life; now they had to give up their freedom if the Communist party superiors demanded it.

It appeared to me that anti-war Americans did not want to be confused with facts distinguished by those of us who were there experiencing the war. Although, I have heard Vietnam veterans that said they ran into little of the acrimony and bitterness that is talked about, I for one that I *did* encounter terrible things. I was called a "baby killer," a "murderer," and many other names that I won't glorify by writing them here. I do know many veterans returning encountered the anti-war sentiment personally.

As presidents and historians have noted, the Vietnam war was the most misunderstood war Americans had ever fought in. Once rebellion and anti-war sentiment became part of the national tone, the media stoked the flames. Details that might have turned the tide and supported the war were either misinterpreted or purposely slanted to indicate "proof" for anti-war leanings. One event often mentioned as an element that continued to solidify anti-war sentiment already brewing occurred when the media released a photograph of Phan Thi Kim Phuck, later referred to as "The Napalm Girl." She was photographed running from her village naked, burning from a napalm strike. The accompanying report that appeared indicated that Americans had been responsible for the strike. In fact, the misdirected attack was a mistake that had been perpetrated by the South Vietnamese Air Force forces with no Americans near there. There were *no* Americans there, even the AP photographer Nick Ut, who subsequently won a Pulitzer Prize, was Vietnamese.

Another event plagued with misinformation was reporting about the Tet Offensive in 1968. Often described by the press as an event when the North Vietnamese attacked during the Lunar New Year, the TET holiday, in military terms the six-hour battle at the American Embassy was inconsequential. This attack was just one of the surprise attacks by the North Vietnamese Army, and the six-hour battle was a disaster for the North Vietnamese. The Northern Army, as well as the VC, was forced out of South Vietnam. The press and many historians even today call TET of 68 a significant win over the Americans and South Vietnamese. It was a lie then as it is now.

Unfortunately, many people based their anti-Vietnam War beliefs on what those of us fighting the war saw as duplicity of the press. By the time of my tour in Vietnam, many of the troops had learned the hard way that news reports landing on front pages and television screens across American provided the *last* information on the war that should be believed and taken as fact. It's a sad situation when perceptions of those who were there and involved differ so greatly from those whose experience comes from what's being fed them from sources who have an agenda that might or might not be based on the whole truth.

In fact, I was living at home when my father flew in the Vietnam War. I know what it was like to witness the news reports and then receive letters from Dad explaining what had *really* happened. This anomaly also occurred when warriors would send audio tapes home, explaining something they had just witnessed or been involved in. When the news reported it, the world had two different versions covering the same event. Was one true and one false? Was the truth some strange combination of the two? I tended to believe the soldiers on the ground whose experiences were not second or third hand, filtered for propaganda or public sentiment to sell news. I would see these discrepancies in action when I was there and found the damage done to families back home when the news had little resemblance to the facts.

I was stationed in the Washington and Baltimore area, specifically Fort Mead Maryland. In certain areas, the hatred was actively anti-war, especially in the University of Maryland areas. That was fine, but many made no distinction and were anti-military, as well. In my opinion, that was not OK.

Today, the talk and public interest about the Vietnam War has decidedly changed since I served. When I'm wearing a Vietnam veteran hat, I have both men and women and people of all ages, even teenagers, thank me for my service and tell me "Welcome home." I deeply appreciate that. When I wear my hat, and I see other Vietnam vets wearing a similar one, I go over, thank them, and welcome them home. It's also an opportunity to find out where and when they served and just say thanks.

LESSONS LEARNED

1. I learned that I should fly with the most demanding pilot because that person is probably the best teacher for mission specific information. I tried to learn whatever that person knows and is willing to share. This information will make me more efficient and a better pilot.

2. I learned to listen to the hanger flying discussions between the more experienced pilots. They will share little tidbits of information that they learned the hard way. I don't need to relearn lifesaving information they've already learned the hard way.

3. I'm not going to psych myself out by dwelling on the hardest parts of a mission. It will be hard because at its most basic, the enemy is trying to kill us, or at a minimum, hurt us enough to remove us from the fighting. This is known knowledge before I arrived at the war zone, so dwelling

on it is just relearning what I already know. I will dwell on being the best possible pilot I can be because that may save me and my crew's lives.

4. My attitude about being injured or sick may very well be very significant in my recovery. It is a war; people get sick or injured, and those are just facts. If it happens to me, I will learn to live the best possible life I can within the limitations I now have.

INDIVIDUAL FLIGHT RECORD 1971: AUGUST 1971

INDIVIDUAL FLIGHT RECORD AND FLIGHT CERTIFICATE - ARMY (PART II)					26. PL. O COVERED 1971 Aug				27. SHEET NUMBER 3-2			

28. LAST NAME - FIRST NAME - MIDDLE INITIAL
McCARTHY, JOHN M.

29. SERVICE NUMBER SSAN

DG GRADE AND COMPONENT WO1 USAR

SECTION V - FLIGHT HOURS ACCRUED - TOTAL HOURS FLOWN BY MONTH (For Local Use as Desired)

JULY	AUGUST	SEPTEMBER	OCTOBER	NOVEMBER	DECEMBER	JANUARY	FEBRUARY	MARCH	APRIL	MAY	JUNE

SECTION VI - RECORD OF FLYING TIME

					FIRST PILOT FLYING TIME									COPILOT FLYING TIME							
						FIXED WING				ROTARY WING					FIXED WING			ROTARY WING			
							NIGHT				NIGHT					NIGHT			NIGHT		
DATE	AIRCRAFT TYPE MODEL, SERIES	MISSION SYMBOL	AIRCRAFT COMMANDER	INSTRUCTOR PILOT	FIRST PILOT	WX INST	VFR	WX INST	HOOD	WX INST	VFR	WX INST	HOOD	CO PILOT	WX INST	VFR	WX INST	WX INST	VFR	WX INST	CROSS COUNTRY
a	b	c	d	e	f	g	h	i	j	k	l	m	n	o	p	q	r	s	t	u	v
Aug																					
2	UH1H	C			7.0																X
4	UH1H	C			1.3																
5	UH1M	C			7.0																L
7	UH1M	C			6.6																X
11	UH1M	T			1.3																X
12	UH1M	T			3.3																X
17	UH1M	C			3.6																X
18	UH1M	C			3.3																X
20	UH1M	C			1.5						.5										X
21	UH1M	C			3.5																X
23	UH1M	C			1.0																X
25	UH1N	C			5.5							.5										H
26	UH1M	C			2.5																X
28	UH1M	C			1.5																
31	UH1M	C			4.0		LAST ENTRY															X
1. TOTALS THIS SHEET					53						1		1								
2. TOTAL BROUGHT FORWARD FROM SHEET NO 3-1					168						8	1	3	39							
TOTALS TO DATE					221						9	1	4	39							

A GUNSHIP MISSION

Darkness was just becoming light. It was early—about 0600. This was somewhat early for a non-emergency flight, but I was excited to fly. As we started to back out of the revetment, the AC told me to take over the radios. I wasn't too concerned about this, but if we entered a hot situation, the radios would be busy. I was pleased that the AC, who we called JB, had so much confidence in me.

JB was our senior line pilot who had found his niche in life as a gun pilot. He had completed his first tour and was finishing his second; however, he had extended for another six months. Exceptional at his job, he was in his happy place. This wasn't only my opinion. All the fun ACs thought he was the best they had ever seen. To have him trust me, an FNG co-pilot gave me confidence in my progression.

I called us out to Dong BA Thin tower: "DBT tower Sidekick 7 us a flight of two guns ready to move to the departure area."

DBT responds, "Roger winds are calm, reposition direction

of departure your choice. No other aircraft reported in the area. Advise when ready for departure."

We lined up heading north in the lead, and Sidekick 3 lined up behind us. As Sidekick 3 called, he was in position, so we backed out of the revetment into position. JB called, "Takeoff checklist complete." We were heavy as we flew to the staging area, loaded with ammo including fourteen, seventeen pound warhead rockets. It was possible we'd bounce off the tarmac during takeoff, so we were backed all the way to the wire.

I called us out to DBT tower, "Sidekick and team are departing to the North. Guns are cold and on the roll."

We cleared the fence and had started a relatively slow limb. Sidekick 3 was the wingman, the second aircraft called up in formation. In about three minutes, I rogered that we saw an field. We never saw anyone, so we used it to test fire the guns. JB told me to fire a three-second burst into the field, and it all worked. Then he told the crew to fire their 60s in a thr4ee-second burst, and that all worked. He then instructed Sidekick 3 to do the same, and all went well. I called us clear of DBT airspace, and we changed frequencies.

We were going up the coast, so we headed to the east to the South China Sea. I called Nha Trang tower and told them where we were. We were transitioning through the area "feet wet," which meant we were going low level over the sea where a lot sampans were.

We had caught one with Ak 47s that we could see. We received clearance to sink the boat, leaving the passengers to swim. We couldn't fire on them since they made no move to shoot at us. We could only hope for a repeat if they decided to grab their weapons, but that didn't happen. When we went feet wet, we were under the ridgeline that was part of a large plateau where the airbase was located. We were just above the boats.

Nha Trang Airbase had fast movers flying in and out, and

we didn't want to be in the idle of them, plus low-leveling responsibly over the water was good. When we called clear, we went inbound to the base of the mountains.

On a previous flight, we were climbing up to clear the cliff when I saw what looked like a pillbox, a concrete structure from which troops would shoot at boats and people disembarking. I didn't know that during World War II, Japan had occupied Vietnam and had defense system from that time.

JB knew where we were heading since he had staged from here a couple of times. As we flew closer, we all amped up a notch because we were in bad guy country with plans to run then out. This wasn't supposed to be a big operation, but it did have a lot of secrecy to it. We were working with a company that flew out of Pleiku, so I didn't know them. We were talking to the ground unit securing the LZ to find out what they wanted from us and where to land. At this point, I was able to look around more to get a better feel for the area. I had already learned when things go bad, as they sometimes did, I had better know the terrain and be prepared to use that knowledge.

We landed at the designated area, and we walked up to the pilots gathered for a mission briefing. I knew this mission was different because we were given no indication what it was going to be. On the surface, it seemed to be a standard insertion mission until the end of the briefing. Intelligence was available that a large force of NVA were making their way to this area for some unknown reason. We also learned that a smaller group of grunts were involved because a larger group might have tipped off the enemy, and they would go underground for a while. The ARVN intelligence provided the information, and they were usually reliable. True or not, we would find out shortly.

We all returned to our aircraft for one more smoke and a cup of coffee we had brought with us. We talked about the mission and decided we all felt ready. We knew from the briefing that if it went hot to a larger force, a resupply of weapons and MedEvacs would be coming. In this case, it would be a very hot operation.

We received the start up signal, so startup we did. We moved into position and deprted slightly ahead of the slicks in the event the enemy was close. If they shot at us, we'd break them from the habit. We were on the outside of the slicks, and when they started their descent, we went to work. Looking out of the windscreen, all we could see was some open areas in scrub brush. Anyone on the ground would have known that helicopter operations were happening. I suspected some NVA did, so I shot the brush and the real LZs with mini-guns, as did Sidekick 3. We met no ground resistance, so we thought this was another military intelligence miss.

As the slicks turned back to the pickup zone (PZ), we heard some random shots aimed at the grunts. The slicks reloaded grunts, and off we went. We were down to about sixty percent ammo, but we still had rockets, so we were a lethal force. As the aircraft were on short final, a 51 caliber machine gun opened up on them. We saw where they were firing from and engaged. Apparently, we got them. Then a second 51 caliber machine gun opened up about one thousand feet from the original. They weren't hitting the slicks, and we were going to leave them for later, so as to not take away the covering fire we could use for the slicks.

It was an excellent choice because the world came alive with AK-47 fire and B-40 rockets. I hate B-40 rockets because I could see them coming right at me. It was damn near impossible to avoid one on the right track. Thank God, we didn't take the bait of the 51 caliber that far away. It would've been a massacre that we were able to avoid. We were sure we had some hits, but our instruments were in the green, and the last thing we wanted was not to go back there. We refueled from a fuel truck that had arrived, and armament people rearmed us. We had virtually nothing left to shoot when we got off station.

We could hear the grunts who were fighting a far numerically superior force, but not a superior fighting force. They were running out of ammo. A call was made from more slicks for a Medevac aircraft for support. A couple of MedEvacs came next, but so many troops needed gun cover that we did the best we could, and they did the rest. They picked up the wounded, and we started taking a toll on the NVA.

Looking out the windscreen as we covered the troop on the ground, I felt like we should be able to see them since there were so damn many of then. We were now having to manage our ammo more because of the sheer need for our support They had called for more gunship covereage, but right then none was available.

This situation was the beginning of the drawdown of American troops, which meant pilots and aircraft were not as plentiful as before. It was early in the drawdown, but we felt the effects and knew it was going to become worse, not better. We went on throughout the day, and it seemed as if the grunts were making headway. By the time the battle started to end, the NVA must have gone underground because we couldn't see them or draw them out to fire at us.

The mission ended with the withdrawal of the ground forces with virtually no further contact with the enemy. I felt like I had been beat. It was physically draining. Gratefully, we didn't have to do a ground rehash because the commanders had a tremendous amount of information to cover, and we could do our recap later. When JB did the recap, he told us what we already knew. The American soldiers fought long and hard and had no type of failures to discuss. In the grand scheme of the war, this was just a hot battle handled well by the American forces. We understood that no American deaths occurred, but a substantial loss of enemy troops did. The numbers weren't over inflated; we saw it and lived it.

As the briefing continued, I discovered that analyzing the battle step by step was helpful and informative. The only way I can describe it when it was happening in real time is surreal, like I was in some dreamscape. I had looked out the window and through the mini-gun sight. Sometimes I saw the enemy, and sometimes I shot where the grunts told us where they were drawing fire. Sometimes I shot where I just thought they could be. At any rate, I knew that I had played a part in this success. It will never be more than a mission flown, but I saw the American fighters in action. Unfortunately, this battle would go unreported by the press.

Looking back, I realized this was the kind of mission that gunships were created for. The outcome of this battle would not have been the same without helicopter gunships. We had done our job, and we had done it well. More American troops went back to base alive because we had been there.

LUCKY SHOT

As a brand-new Aircraft Commander, I worked hard to use and practice everything I had learned in my first four and a half months flying in a war environment. I knew how to get the job done properly; I knew how to stay alive. The Army had made a significant monetary investment to train me to be a combat helicopter pilot. I wanted to make sure that I delivered their money's worth, so I became relentless in doing what I thought was right. Furthermore, I knew how much it meant to me personally to do the job to the best of my ability, especially considering everything I'd done to be in the cockpit flying. Even with all my knowledge and training, I was aware that on any given day, a flight could, and usually did, show me things I had not seen before. Today would be one of those days.

We were set up in a field close to the "Trail." A farmer had reported to the intelligence people that NVA troops were in the area, possibly a lot of enemy troops. A company of our infantry troops were sweeping it. A truck had dropped them on the road close to the trail, but since the distance from us was a couple of miles south, we had no contact with these grunts. Proper communication is an integral part of a successful mission, so we were a little on edge about our location in the middle of this field with no protection or reinforcements. This operation could easily go south on us. I started to have an uneasy feeling.

About 10:30 a.m. we heard the first words from the soldiers on the ground. They made a quick, quiet call on the frequency we had been assigned to use on this mission. We had a PRC 25 radio for our communication with them. They had seen some troubling signs and recent activity of a good-sized number of enemy troops walking in the brush and tree line next to the track they were on.

Soon we heard from them again in a noticeably quiet radio call that they had seen a couple of NVA just ahead of them along a path that included parts of their preferred track. They had not engaged these two people in a firefight. Their plan was to follow them to see if they were joining a larger group.

About fifteen minutes later, they requested us to prepare the aircraft to launch. Things began to happen fast. We knew that our people numbered about fourteen since two squads had been assigned this mission. The remaining two squads of the company were standing by a couple of miles south of us. We received a radio call that they were waiting on some aircraft to fly them up to our staging field. To our way of thinking, the part of the plan that involved us didn't feel like a good way to do things. None of the plan that would now be executed had been passed on to us, even though we were directly involved. As the aviation assets, we needed to be very much aware and involved in the agenda since we would obviously be covering these places. I didn't like the looks of this.

I noticed that the very hair on the back of my neck began standing up. I had a sudden realization that things may be going bad. Just minutes later, we could hear the blasts of firefight as we prepared to go. We had to be ready the second they called us. As I listened, it occurred to me that perhaps the fight had been going well for these first few minutes. I breathed a sigh of relief but didn't let down my guard. Then the call came in.

As the radio blasted the message for us to launch, loud voices shouted and barked the commands, breaking the previous calm and quiet. The enemy was growing in numbers, but the grunts didn't know where they were coming from into the fight. The constant swarm of boots on the ground seemed to be appearing out of nowhere.

As we began launching, everything went like clockwork. We had a plan, and we knew how to execute it. We set up for our attack pattern, and within a minute we started taking moderate fire that was coming from the direction near the engagement. Our Lead Pilot broke us from our planned route, and we went into a standard attack oval pattern with one gun inbound and the other one flying and climbing on the outbound leg. The crew chief was covering the inbound aircraft with his M-60 machine gun while the gunner on the left side of the aircraft was shooting down into the brush and trees hoping to get some of the enemy troops heading to the fight.

The grunts were becoming more excited by the fact that this superior number of enemy troops was ever increasing. By then, we had figured out that the NVC must be coming up through a maze of underground tunnels. The grunts hadn't reported seeing any holes or tunnel entrances at this point, but they couldn't see any signs of vehicles either. On the ground, they didn't know how it could be that bigger and bigger numbers of enemy troops continued to head this way. The tunnels were the only logical explanation, so we embraced that knowledge and proceeded with deliberation.

These holes in the ground created a protected tunnel-network packed full of the enemy. They poured out of the entrance like ants, creating a significant and immediate danger to everyone involved in this mission. Our Lead communicated with our operations people to tell them to put aircraft into this developing battle because it was clear that we would be facing more enemy troops than originally expected. We were also going to need fuel, rearmament supplies, and at least one more team of guns and slicks in case troops needed to be extracted.

This battle was fast becoming a bad nightmare, growing more terrible by the moment as the numbers of the enemy grew rapidly. In the midst of the smoke and noise, I could tell that the battle was becoming more and more intense. Bullets were flying so close and so rapidly that I could hear them zip across like buzzing bees. From my perspective, I could see through the smoke-filled air that ever-growing number of troops shooting at us.

As for the outnumbered ground troops, the numbers of casualties had begun to mount. In the white-hot moment, I could see the awful truth that the number of our troops was diminishing through attrition. Since the Lead ordered the beleaguered men out of the battle zone, we were able to go after the largest enemy group without endangering our troops on the ground. It proved to be a lucky move to withdraw the ground troops because not only were they outnumbered, but they were also close to being out of ammunition. Staying would've cost more American troop lives.

Suddenly, a group trucks carrying ARVN troops arrived, bringing ammunition for the grunts and rockets and ammo for us. In our last gun run we had fired all the ammo we had left. Presently, we were turning outbound to rearm and refuel.

Then we heard it. "Boom! Boom! Boom! D'Boom! D'Boom! D'Boom!" A series of the loudest thunderclaps I'd ever heard, growing ever distant until they progressed further and further away. What began as an earth-shattering explosion finally diminished to nothing. The trail and area around us were shrouded in smoke.

In the aftermath, we did not know exactly what had happened, but some of the grunts on the ground had a theory. The last round we fired must have been a direct hit into the rabbit hole tunnel entrance. Since the enemy was known to store ammo underground, the explosions we heard may have been that ammo going off in a series. Suddenly, what we had heard made sense to me. I'm sure that the NVA were storing huge amounts of ammunition for their next big battle to take over this part of Vietnam. Logically, this situation would require enormous amounts of ammo, and some would have had to be stored underground for easy access during a major attack.

Of course, we will never know for certain. It could have been one of the last rockets we fired, or it could have been a screwup by one of the NVC fighters trying to get out of the tunnel to join the fight. Whatever caused the explosions, the battle ended immediately. We waited until the explosions stopped, and the commanders could get troops in there to do a bomb damage assessment (BDA). While we waited for whatever would come next, we talked about what we each knew or had seen in the time leading up to the explosions. The more we all talked it out, we started to believe in the real possibility that one of our rockets shot right down the hole had started the explosions.

That night as I tried to go to sleep for a couple of hours' rest, I mulled this over in my mind. I found it hard to believe that I had accidentally hit that tunnel entrance without even seeing it. The tunnel entrances were small, spider holes small. The number of people killed by those explosions on that day had to be a large number. For now, I would leave this alone. I did not want people to believe that because I had hit that spider hole with a lucky shot that I was interested in earning credit for those kills .... if, in fact, it was caused by one of my rockets.

About a month later one member of the Military Assistance Command Vietnam (MACV) came to visit with us at Ban Me

Thout at the tiny wooden club we had. To call it a "club" stretches the mental picture. We had a small, wooden building with no running water, only marginal electricity, and no refrigerators or cooking devices. It was a place to sit down and talk. The MACV was an interesting person. He had been seriously wounded during his last tour in Vietnam. He pulled his shirt up and showed us where he had been shot with what appeared to be five bullet holes around his belly button. He had spent a long time in the hospital going for surgeries and recovering but volunteered to come back to Viet Nam for another tour. As a Military Assistance Command Officer, his role was to help integrate the ARVN soldiers into the American way of fighting in a war, a truly thankless job. These people were serious, as a rule; however, he had a good sense of humor and recognized the value of helicopters in a war and the role of gunships. A helicopter gunship with a sharp-shooting pilot had saved his ass in more than one situation.

He had just finished working with the grunts and the forces he integrated with them. In the last couple of minutes, he turned to us and asked about the incident with the underground tunnels blowing up. He got a resounding answer to what happened: one of the last two rockets hit the spider hole directly. Pulling a bag of hand grenades and other explosives, one NVA must have been coming out of the hole just at that moment of direct hit. The collective group in the club believed the NVA was killed by the rocket and then released the bag, which might have armed a grenade causing the explosion of stored ammunition where the spider hole was. This was the "house that Jack built" theory only. We will never know for sure.

Also, a captured soldier trying to escape through another hole shared the information that this tunnel location was a huge operation underground, which included a surgery area and a small medical clinic for their wounded troops. Everything and everybody were most likely destroyed by the blast event. It may have just had a huge wall of dirt that blocked the ammo from the clinic, but whatever the arrangement, some careless handling by the NVA underground occurred.

As aviators, we had not received any advanced information about this massive underground complex that had been so important to the enemy, but we had everything to do with its destruction. We had a distinctive impact on enemy plans for this area of Vietnam. Whatever way we chose to imagine it and the consequences, that was one lucky shot down that spider hole.

LEARNING ON THE JOB

The days wore on. I flew a heavy amount since I volunteered with the slicks when I wasn't scheduled with the guns. At night we mostly sat around our own bar and told flying stories, a situation called "hanger flying." This shouldn't imply that our talk was about extraneous bullshit. On the contrary, the A/C's dissected our actions as a team and created a personal building opportunity.

Sometimes I learned something that it was easy to forget: that the pilot on the right (the Right) had the opportunity to learn something. The perfect example of this was when we were inbound on a gun run. This was an active situation to say the least, and I was at the controls when the AC put his left hand over the transmission and engine oil pressure gauges, demanding that I tell him what they were reading. My reaction was to shout out immediately, "I do not know, but they're in the green." This did not satisfy him. I remained as calm as I could be, but I did not match his level of calm, as he set about

149

flying, fighting and teaching because he wanted to explain that it didn't matter how busy I was. I should have a crosscheck that included the aircraft operation and the mission part. We were taking fire and returning fire during this discourse.

"Sir, can't we address this later?" I asked. That's when he explained that this very situation was what he was talking about. I should have the ability to listen and talk both inside of the aircraft and outside on the radios, all while keeping up with the action on the ground and in the air. He stopped, and in between runs there was shooting the mini guns at active targets I observed him much closer than I had before. He was flying, shooting rockets with amazing accuracy, talking on the radios, and maintaining proper control of the situation outside and inside the aircraft. He was also a "high school to flight school" pilot, one of those who some thought could not fly with the best. It did not matter what degree with which you ended your college career. It could be something as removed as Forest Preservation, but the belief was that you had to have a college degree to have the knowledge and drive to learn more. So here was the example of the very people often excluded from flying with no good reason. In addition, to gain admission to Army Officer Candidate School (OCS) and then Flight school, you had to have an IQ of at least 125 while to enter the Warrant Officer Rotary-Wing Aviator Candidate Program (WORWAC), it had to be at least 150.

The fight was coming to an end as the effects of our firing took its toll, and the remaining enemy left the immediate area. The grunts had been dropped off by the slicks. They were in hot pursuit. We were not shooting, but the weapons systems stayed hot in the event we had to cover any of the aircraft as they were reforming to base (RTB).

At the end of that mission, we were released and headed to the revetments (physical barriers on either side of the aircraft) in parking. We refueled and re-armed and went on to parking.

We were post-flighting the AIC when he walked around with me asking valid questions as we looked for bullet holes or other damage that may have occurred during the mission. It amazed me that for as much lead as the bad guys were shooting at us, so few bullet holes resulted. The situation this time didn't diminish the fact that sometimes a lot of damage occurs, including the shelling of a crew member. As we talked about the mission, I could tell he had something important to say to me. Naturally, I assumed I was going to be questioned about my answers during this episode. Instead, he wanted to explain while we were now in a calmer environment, why today had given him the opportunity to teach me such a valuable lesson. He explained to me the obvious: that any pilot shot down and captured was going to go through hell by the captors. Gun pilots were exposed to a particular kind hatred by the enemy; they often believed the pilot captured was the very one who had killed some of his family and friends. The possibility of being recovered alive was improbable at best. That aircraft was my only possible way of escaping the immediate area if I was shot. I would have to fly it until I hit the ground. The few seconds I might be able to give myself would be directly linked to the instruments that warn the pilot that the flight will terminate well short of the destination.

When one of us went down, the other gun would fly through anything humanly possible and beyond to keep the enemy away. I had lessons in survival that played a huge part of my training. As the saying goes, they will "fly till they die" trying to protect you. Suddenly, I realized this man was a walking encyclopedia of how to ensure the biggest edge to survival, while doing the mission beyond the expected.

All the yelling and screaming in flight school now became clear as a bright sunny day. Keep your cool, remain focused, and fight until you have no more fights left in you. Then fight some more. Having been the cover bird on a downed gunship more than once with this AC, he had taught me such valuable insights.

There was no ceremony when it was time for us all to pay our respects and honor our fallen heroes. When the recording of Taps began, started, I could barely see for the tears, and I had to go stand outside. Soldiers died fighting to give the South Vietnamese people the right to vote in fair and honest elections and determine their own destinies. As the Bible says, "Greater love hath no man than this, that a man lay down his life for his friends" (John 15:13). As I type this fifty years and a few months later, I still cannot think of these men, and many others I have known, without having to struggle to hold back a sadness and overall sick feeling and wondering when man is going to find a way to end war.

MONSOON SEASON

A monsoon is traditionally a seasonal reversing wind accompanied by corresponding changes in precipitation. Usually, the term is used to refer to the rainy season. The phrase is also used to describe locally heavy but short-term rains. The rains in 1971 were very, very strong yielding a lot of water. The monsoons caused major damage to structures and flooding, but in addition along came a major typhoon—Typhoon Hester. The typhoon caused considerable damage and even disrupted the war. The storm landfall was over the military installation of Chu Lai.

Originally the storm was to hit right over our base. We prepared the aircraft to move to a soon to be determined base, so we waited. As the storm was approaching, it was obvious that it would hit over the top of Chu Lai, so people were evacuated down into our area. As we prepared the base for the storm, with very little time to prepare for visitors, a helicopter company from the Americal Division came to our base to reduce the damage to their aircraft. We were ill-suited to absorb that many more aircraft or room for these troops to bunk. Even in this crowded and uncomfortable situation with an impending storm, we somehow made it work. We had the visiting pilots sleeping on our furniture and on the floors in sleeping bags. Overall, it was good having pilots from a different base. It gave us an opportunity to learn about their threats and tactics to keep an upper hand in their district.

With typical Army "can do" attitudes, we made it all work out. The rainfall was more rain than I had ever seen come down at one time. Picture relentless sheets of rain, buckets of rain, torrents of rain. Add to that incredible volume of rain strong winds blowing so hard the trees bent and indiscernible things unattached well flew away. Somehow, we made it through with no real damage to the aircraft or to any troops.

Chu Lai, however, was heavily damaged. Due to the damage, the aircraft and crews stayed over a couple of days, so emergency repairs could be done. There were at least 100 people killed by this typhoon in Vietnam and more in Laos and Cambodia according to some newspapers. On October 25, a Republic of Vietnam air force plane crashed causing thirty-three fatalities. Three Americans were killed by flying debris and twenty-one others were injured. Thunderstorms created by Typhoon Hester caused a U.S. Army helicopter to crash killing ten more Americans.

The time dragged by when the monsoon storms hit. It rained harder than I had ever seen. We had precious little to do except play solitaire. Being gun pilots we had to be ready to go, even ready when the rain was so hard you could not see through it. Due to the exceptionally high humidity and heavy rain leaving our boots on, which we had to do, could and did cause us to get what was referred to as "jungle rot." This was not normal for aviators, but these were different times. It was unavoidable for our boots not to get wet. We couldn't take them off to dry our feet and let our boots dry since we had to be constantly ready. That constantly and unrelenting wet environment started the problem, and it did not go away easily. Even as aviators we had to deal with a ground-pounder's problem.

Every now and then the rain would stop, and it might be possible to fly, so we would go fly and do our job. Of course, our own experiences taught us to sit down and relax when the weather made it impossible to fly. It was hard, but if there was a request for us and the weather was so bad it was unsafe to fly, we turned it down and kept our thoughts on those who needed us. Daily throughout the storm, we hoped and prayed that the weather would break even if only for a short time to enable us to go. When those flights did happen, and we took them, they required some excellent skills because we had to fly very low and often so low that we had to make sure we didn't shoot ourselves down when firing rockets. This also required a slightly different skillset for firing because the angles were not there. It was just shooting basically level and at tree top level. It was challenging work and often demanded all you had in your skills and situational awareness plus monitoring radios. It was like what we did normally, but at an even lower level while working to be effective for the ground troops involved and for ourselves.

Back at the rooms, we had to put sandbags in front of the door to keep the rain from soaking everything inside. Boredom set in. It's always a real threat to the warfighters, causing everyone to think that this rain-drenched land was not worth the cost to hold on to it. This attitude ate through the fighters' drive and required some serious headwork to get back into the war.

During the monsoons we encountered these odd shafts of rain coming down. You could fly around them and make headway. This was not an aviator's dream, and it was incredibly stressful. You could see the shafts of pouring down rain ahead, but you could not actually see what was right in front of it, so the flying was difficult. With things on the ground being so boring, any flight would be worth the risk. That feeling was not successful to higher ups.

There was an amazing amount of rain in Vietnam, and the monsoons add a huge amount of rain every year. The area I flew in was mountainous and didn't have a lot of rice paddies, a lot of rain still fell. By the time the monsoon season slowed down and ended this unit was getting set up to shut down, and that happened.

UNUSUAL EVENTS ON THE BATTLEFIELD

Generally, very little humor exists on the battlefield, and if any does occur, the variety will often be dark humor. As we know, military intelligence is always on duty and can uncover plots and plans made by the enemy. How this is done is basically a secret although one that isn't that hard to figure out on occasion.

We were sent out for gun cover on an insertion that should have been relatively calm. As far as we knew, no large numbers of enemy soldiers occupied the area right then. We all did our jobs getting ready for this operation. We had heard that an English proficient soldier was in the ranks of the NVA in the area. Some of the other pilots had heard him on the radio and said he was sounded quite convincing, so we were warned not to be led astray by this man.

The NVA had a good setup to trap some of our aircraft and troops, but one major thing was overlooked. The slicks knew where the LZ's were to be, but when we arrived, we received a call that the LZ had to change because of enemy activity. The caller came up with some coordinates and passed them on to us, a huge clue that this was not right. On the VHF radios, the slicks were talking to us about how to cross up the caller perpetrating the trap. I heard all the calls, including ones changing things, and I surmised that the ground leader didn't sound exactly right. Some pathfinders on the ground setting things up caught on, also.

We knew that if, indeed, this was an English speaking NVA talking to us, they would pop some color other than red smoke, even if we told them to pop *red* smoke. Red smoke meant that the LZ was not usable; apparently, the enemy did not know this. To shift to the color communication to the color red that we wanted, we went back to some of the old notifications, such as "What color uniforms do the Baltimore Colts wear?" We knew the NVA may not know the answer, but we believed this English speaker was IVY League trained, so he may know the colors for some of the northern teams. What we really wanted to see was how much red smoke popped up. So, we called for red smoke to mark the known LZ's. When we did there were about six red smokes that popped, and we now knew our target locations. I don't think the secondary colors for the Colts were necessarily relevant, but it was a way to cross them up.

As soon as the red smoke showed, we went to work. In the meantime, the slicks were mouthing off to this English speaker about how stupid he was and how all the carnage was on him, hoping the NVA would make him pay dearly for his mistake. We don't know what they did to him, but he was not heard from again.

The outcome was kind of funny, but it could have been very tragic had we not known he could be there and would try to alter our plans. That prospect could have put us right where they wanted us. We would have been taken out of the war. Our situational awareness was required to the utmost, and an

English-speaking, Ivy League NVA was part of one of the plans to take us out.

One of the other situations we had involved a Soviet Commander on the ground with some of the better NVA companies. Some of the pilots saw him very clearly on occasion; he didn't really try to hide. I believe I saw him one time for a couple of seconds. We were receiving some heavy automatic weapons fire, and as I made my break to the outbound leg, I caught a quick look at a Caucasian man in a different uniform. The sighting was quick, so we couldn't do much about it, but I was pretty sure I had seen this Russian.

One of the other oddities I had seen was while we were receiving moderate fire in one of the mountain passes that our vehicles used to get up to Banh Me Thuit. It was usually active, and I could not figure out why so many enemy troops landed in this pass. After I had been there for about six months, I found out why.

We had been in a fight in the pass with an NVA force when suddenly they just broke off the fight and disappeared into the jungle. We had fuel and ammo enough to go after them, and that we did. We flew to the end of the valley we were following and popped up over the hill. There we saw what must have been 100 NVA troops swimming in a very well-hidden mountain pond.

In all the times I had been there, I had never seen this pond or realized this was an R&R center for the enemy troops. We fought them until we were out of ammo and had to depart and rearm and refuel. As soon as we were refueled and had a fresh load of ammunition, we took off back to that same area. When we arrived this time, we saw no one—not one person. Not one NVA or VC, and we looked very thoroughly. We figured out they must have had tunnels up there and hid out whenever American troops were in the area.

One of the more unusual events for us had to do with

rockets. We had been in some heavy fighting, and we were expending rockets like we had a never-ending supply, which we usually did. This time, however, we had a real shortage until we had none. The ammo bunker was empty of rockets. We had plenty of machine gun/mini gun ammo, so we could still fight. Our ammunition guru called the Air Force, and he secured a thousand rockets from them. A thousand is a very large number, but the Air Force system worked that way.

We had several loads of deuce and a large truck-full of them. We had to take each rocket out of the cardboard tube and screw the warhead on. This chore was not overly difficult, but it was time consuming.

Luckily, I was called out to join a team we had flying to make it a "heavy team," so we got out of that tedious detail. When everything the "heavy team" completed was over, and we returned, we were pumped up. The fighting was so severe because we had no rockets, but we persevered, and we prevailed.

We refueled and rearmed, and the commander decided we needed to test fire the Air Force rockets, so off I went. I fired a pair of rockets at a tree and had a direct hit, so all was good, I thought. We called this in, and they asked for me to shoot all seven pairs to make sure, so I lined up on another tree and fired. One rocket went left, and one went straight ahead. I thought I knew what happened, so I radioed back and told them to check and see if anyone dropped one of these rockets. No one owned up to it, but I knew if you dropped the rocket, you could break the solid fuel. It was gooey and not so much solid like a brick, more semi-solid.

I turned around to fire again, and I had a perfect pair. Next, I made one more run in and had them both go in different directions. I didn't understand what happened, so I went back to base and found somebody I could talk to about the rockets.

To my surprise, I was accused of being out of trim and

flying too slow. They sent JB, another pilot, to run a test, and when he returned, he was perplexed. He explained that I had done everything right because it happened to him. We all stood around looking at these rockets when one of the crew chiefs told us to look at the exhaust back where the fins were. They looked perfectly straight. He explained to us that the slow-mover rockets had a flanged exhaust to create the spinning motion that was required because speed was needed over the fins to keep the rocket on track. These had no flanged fins because they were used on fast movers, i.e., on the F14.

We learned that evening that we could not return the rockets because the Air Force had no way to guarantee that they had not been tampered with. This meant all the warheads had to be removed, and the rockets moved out of our ammo dump to another revetment to store them where they were relatively safe. We had no use for the rockets since they could not be controlled. Eventually, they were traded off to one of the bases for something we could use. This method of bartering was just the way the system worked.

LESSONS LEARNED

1. I will always remember there is a true dichotomy of life in a war zone. On the one hand, I looked forward to days off, "down days," when I was flying a lot only to then look forward to being back in the cockpit when I wasn't flying.

2. I learned that things can change in a moment's notice. At times not much was happening. It was fine to pass the time playing cards or writing home or just sitting around with my fellow pilots, talking about whatever we could think of. When the VC started shooting at our rooms during a heavy rain event, we all went to the fence line and shot back at them until they had enough of us. Then boredom was back.

3. I learned to be ready and that sometimes all we could do was wait. Of course, we didn't like having to wait due to things beyond our control. I did come to realize that the monsoon gave us a break while allowing for maintenance

to be performed on the aircraft. Still, those wet feet and intense boredom made dealing with all the relentless rain hard.

INDIVIDUAL FLIGHT RECORD 1971: SEPTEMBER 1971

INDIVIDUAL FLIGHT RECORD AND FLIGHT CERTIFICATE - ARMY (PART II) For use of this form, see AR 95-64, the proponent agency is Office of the Assistant Chief of Staff for Force Development	26. PERIOD COVERED 1971: SEP	27. SHEET NUMBER 3-3
28. LAST NAME - FIRST NAME - MIDDLE INITIAL McCARTHY JOHN M.	29. SERVICE NUMBER SSAN	30. GRADE AND COMPONENT WO1 USAR

SECTION V - FLIGHT HOURS ACCRUED - TOTAL HOURS FLOWN BY MONTH *(For Local Use as Desired)*

JULY	AUGUST	SEPTEMBER	OCTOBER	NOVEMBER	DECEMBER	JANUARY	FEBRUARY	MARCH	APRIL	MAY	JUNE

SECTION VI - RECORD OF FLYING TIME

			FIRST PILOT FLYING TIME											COPILOT FLYING TIME							
					FIXED WING		ROTARY WING						FIXED WING		ROTARY WING				
						NIGHT		NIGHT						NIGHT		NIGHT			
DATE	AIRCRAFT TYPE MODEL SERIES	MISSION SYMBOL	AIRCRAFT COMMANDER	INSTRUCTOR PILOT	FIRST PILOT	WX INST	VFR	WX INST	HOOD	WX INST	VFR	WX INST	HOOD	CO PILOT	WX INST	VFR	WX INST	WX INST	VFR	WX INST	CROSS COUNTRY
a	b	c	d	e	f	g	h	i	j	k	l	m	n	o	p	q	r	s	t	u	v
SEP																					
1	UH-1M	C			6.3																I
4	UH-1M	C			1.3																N
9	UH-1M	C			7.3					1.3											I
10	UH-1M	C			1.3																I
12	UH-1M	C			1.5																I
13	UH-1M	C			1.3																I
14	UH-1M	C			2.3																I
15	UH-1M	C			4.0																I
20	UH-1M	C	2.0		0.8									0.5	0.5						H
21	UH-1M	C	2.8																		I
22	UH-1M	C	3.8																		I
23	UH-1M	C			1.8																-
24	UH-1M	C			1.0																-
25	UH-1H	C	5.5																		I
26	UH-1H	C			6.5																I
27	UH-1H	C			2.1																I
28	UH-1H	C			1.0					1.0											N
30	UH-1H	C			1.5																I
								LAST ENTRY													

31 TOTALS THIS SHEET	14½	40					.2	1	1									
32 TOTAL BROUGHT FORWARD FROM SHEET NO 3-2		221					9	1	4	39								
33 TOTALS TO DATE	14	261					11	1	5	40								

DA FORM 759-1 *(PART II)*

INSIDE VIEW OF A GUNSHIP OPERATION

As we readied for takeoff, we all had our own way of preparing. After the checklists are done, the radios call our ready for departure status. Most lead aircraft commanders choose to make the majority of the radio calls, but some don't. Some have the wing or tail aircraft make the tower or calls for departure because they can see the whole flight; whereas, the lead aircraft has no real way to see the team aircraft short of setting the aircraft down 90 degrees off center.

So as the tail aircraft, we make the call. It sounded *something* like this: "Dong ba Tinh Tower Sidekick 4 is a flight of 2 guns ready for departure."

Then the tower would generally say, "Roger Sidekick 4 and flight you are cleared for departure. Winds are 360 at 4, and the altimeter is 3012."

Sidekick 4 replies, "Guns are cold rolling." No talk would follow for a couple of minutes. Then usually a call from tower might sound like, "Sidekick4 and flight you are cleared out of DBT airspace no reported enemy activity in your immediate area and good hunting."

"Sidekick 4 Roger cleared of DBT will be frequency change to tactical, and guns are hot will talk to you on the return."

When the aircraft is a couple of miles out, the weapons' circuit breakers were pushed *in* to make the guns hot. Generally, if we are in our operational airspace, we had certain points away from troops or civilians, and we would test fire the weapons. Failing to shoot before coming into contact was not a good plan since we wanted to know the weapons systems were working correctly.

The gunships went inbound at the enemy and often into the enemy fire, so a non-functioning gun was nothing but a paperweight. When we went inbound, we had basically one goal, and that was to have the enemy fire directed at us and away from the other aircraft or troops on the ground.

I have been asked what was on my mind, and what did I see when it went hot like that? We often could see tracers from any of the semi or fully automatic weapons, both theirs and ours, and we would go after the origin of the green tracers.

At the same time, we were monitoring the radios, working out what the others were seeing and figuring out how much fire they were taking and from where. Then we'd work out our gun patterns to suit the enemy locations and the terrain. One gunship would be inbound while the second aircraft was getting in position to cover the firing aircraft turn. The turn outbound left the belly of the aircraft totally exposed and in a position where the door guns could not cover themselves very well. So, the outbound aircraft rolled inbound just slightly before the inbound aircraft turned outbound. Essentially it was a delicate dance, and the teams worked well together. The cost of a mistake could be deadly to our people. It is a huge responsibility that we took seriously, as seriously as our own deaths, or facing the possibilities of severe injuries, or worse case—potentially being captured.

Not one in the gun platoon failed realize that we were betting our lives that we would outfight the enemy and protect our own people, ground, and air troops, as well as ourselves. While all of this was going on, we were still flying the aircraft, a primary responsibility, and monitoring the system instruments in the aircraft. We were also monitoring the movement of US troops, and the VC, so we could alter our gun runs to be in the best fighting position.

Included with this was whatever necessary conversation was going on inside the aircraft to keep all of us as totally zoned in as we could be. Things happen at a blistering pace once contact was established with the enemy, and the entire crew had to be in the loop during this. When things go to full-fledged battle on the battlefield, no time exists for excess non-imperative conversations. So, from the front seats, we're looking at or for targets while firing what weapons the aircraft commander ordered. Many times, we could not really see our targets, but then someone would find them and express where they were. Taking them on was not a choice; it was the job, and many people relied on each one of us doing our jobs at the highest level possible.

In many cases, villager populations were situated in the middle of these battles, and we had no allowances for collateral damage. Anyone shooting and hitting friendlies would experience a one-time event; they would be removed from duty, and more than likely, an investigation would begin almost at once. Should anyone be found guilty of any of the things a troop could be court-martialed for, life would become nothing that anybody would like.

American military members train almost constantly when not in battle and there is little excuse for killing or wounding innocent people, in most cases. Nothing is a one size fits all answer, and this is one of those times.

In the heat of battle, especially from the air you cannot see behind every little embankment or behind the trees. As people get scared, they will run, and you have but a split second to figure out who or what they are. Besides the complexity of flying a helicopter, flying one into the enemy fire while separating the enemy from the innocents is not always easy. Now factor in three radios, UHF, VHF, and FM, blaring away with the sounds of violent fighting, and soldiers calling for covering firing, and maybe calls for medevac, and aircraft calling for gunship support because they are taking heavy fire. The stakes are the highest in our efforts to save people's lives and stop invaders, and yes, that is what the enemy is. No one will argue that the Communists invaded this South Vietnam to overthrow a government that was voted in, so that they could install Communism. We had a mission; I had a mission as a gunship pilot to do a job.

With good planning, excellent training, and experience as a co-pilot, the pilots have a shot to survive and save many other people. As one of the pilots, I would fly back to base to do the refueling and rearming and post-flighting. Then many times in a hectic fight like this you go and de-brief, what was right and what was not so right, or out and out wrong. Your day is done. Maybe. Except for the possibility of being mortared. Or the possibility of being rocketed. Or the possibility of sappers trying to get into the compound through the wire. Then we're up to go and fight some more. A pilot does not get to stay in a weaponized gunship if they cannot live in this world. Failure results possibly in demotion to co-pilot or being removed from the gun platoon. Same for the other jobs done by these aircraft in the war. In this new type of warfare, the aircraft is close to the ground, and in this type of war, always close to the enemy. If they can shoot you down or shoot the aircraft up wounding and or killing the crew, this is winning for them. If you happen to land or crash where they can recover or kill you on the ground, this is a very major win for them. There is no time for hesitation. You must know what to do in these situations or be able to figure it out fast. On top of being low and at a fast speed for these altitudes, you may also be flying over our troops. In case of aircraft failure of being shot down, you're near enough to ground troops to cause death or burning or major trauma. This is the nightmare of low-level flying in a combat zone. We were all always aware and tried to get away from ground personnel in case of some kind of forced landing.

TOUGH DECISIONS

We were in a fight that was winding down. We thought that we had come through it in one piece, but we would not know for sure until we landed and post-flighted the aircraft. All systems were green, and we were pulling out of our last inbound run on this one. It had been busy at the start. The NVA had put up a good fight but had made a couple of tactical errors that allowed us to reduce their size of force advantage with some good shooting. The grunts had apparently stumbled onto this NVA force that was at least a company size, very possibly bigger.

They were unknown to any of our troops, so they had come across the Cambodian border probably the night before, but we did not know the reason. Luckily, we had been put on standby earlier when the squad leader had seen signs of personnel movement in the grass and weeds. This was a soldier that knew what he was doing. Putting us on standby locked us down, and we had that feeling that we would be called in. Just minutes after we had settled in the aircraft, the call came in for us to launch. We could hear the automatic gunfire and yelling of commands and knew they were in it for sure. We had set the aircraft up for a combat start, so we were ready to go in about a minute and a half.

We launched and got coordinates for their location, and the co-pilot plotted them out. We were close and would be on them in minutes. I have heard that time is of the essence, but this was time that was probably going to save lives. We went up hot and settled into the pattern we would use. We lead rolled in hot while we covered him from where he had gone inbound and got ready. He called taking fire from his 3 o'clock, and we could see the crew chief leaning out firing his M-60 without stopping. Lead rolled out and called his break, and we went inbound. Lead called taking fire from his 9 O'clock, so probably the same shooters. Then we started taking heavy fire from our 3 O'clock, so the gunner lit it up on our left side, and the crew chief was firing under Lead to help them.

We called our break and Lead was turned inbound and let a pair of rockets on our bad guys. The firing from there stopped, and as we went outbound, we found some new players shooting at us from almost underneath us. Lead rolled out and altered his outbound to go straight at these guys while we used the miniguns very effectively on a group of them that thought they had run us off. They were going to help their wounded. Unfortunately for them, we had just altered our racetrack pattern to an oblong pattern.

There were a bunch of them down there, and we had taken a toll on them, but they fired a couple of rocket-propelled grenades (RPGs) at us. I hated RPGs because you could see them coming, but they were so fast, it was almost impossible to escape an accurately fired one. We saw where it was fired from, mainly because he shot it wide of us, and we made them pay for that error.

After that we heard from the grunts that they were on the run apparently in all directions. This meant that they could only chase individuals. We had done a lot of firing, and we wanted to rearm in case there was more of them we didn't know about.

It all settled down, and we headed back to the staging area when the master caution light lit up with an engine chip light. Considering where we just came from, we thought it was probably real. We were a few minutes from our staging area and just a couple of minutes from the last battle zone. According to the -10 (think operator's manual) this light required an emergency landing while preparing for an engine failure. We knew that these NVAs had scattered, and we didn't know if some might be heading our way.

If I didn't do the emergency landing procedure, we could end of with no real choice of landing area except where the aircraft could autorotate to, but if we did land now, we might be landing right into some really pissed off NVA. I had seconds to make the decision, and I chose to continue onto the staging area where maintenance could be flown in to fix it.

It turned out well. The grunts heard me talking to the Lead and they could still see us, so they headed out towards us to clear any of the bad guys out before they could get to us. We landed and shut down, and the crew chief and gunner went to work figuring out what was going on. They had to wait for the engine compartment to cool down, and in the meantime, Lead had called operations, and they had maintenance inbound in another Mike model in case we needed it.

Not long after, the co-pilot and I talked about what went on. I asked him what choice he would have made. He said going against the manual would have been hard for him, but he felt it was the right thing to do.

Another factor he mentioned that I was acutely aware of: we had one of the decoding radios on board, and we had to destroy it even if it meant destroying the aircraft, so the NV could not get that radio. It was a special radio that was on only some aircraft, but it was special in what it could do regarding secure transmissions. That was the wild card for me. I did not want to blow up the aircraft, but we could not take the chance that we could remove it if they caught us. It really was no decision at all.

While we waited, Lead was flying overwatch on us, and before long the grunts came in, and we got to spend a little time talking to them. They worked with helicopters very often but rarely got to meet the pilots, and the same was true for us with them. We thanked them for hanging it all out, for basically running with all of their gear on, and for making noise that some bad guys could have heard, a circumstance that could have been very bad for them. They felt the same for us flying into all of that fire and fighting it out with them. They knew they were outnumbered and outgunned. So, we had a mutually beneficial relationship. That was the most excitement of the day, and we all returned to our base in the evening where we had a couple of beers and dissected the fight, so we could learn something from it, and we all felt we did well. End of the day, or so we thought. A couple of hours later, the mortars started, so off we went. One of our teams caught them and ended that threat for the night. Just another day at the office.

KIDS CAN KILL YOU, TOO

I will not expound on this fact more than to say that some Vietnam kids knew war, and that was all they knew. If their parents happened to adhere to Communism, it stands to reason for their children to act in the same manner as the family leaders.

When we were in Ban Me Thout, we were on an open post, a rarity. This meant you could go off base, a leniency which was strictly prohibited at our home base. Whether this is bad or good is probably debated even today. There are, of course, many dangers in this activity. We only went when there were at least two of us, and we stayed together. We were always aware that we were in the central highlands. These mountains were close to the Ho Chi Minh trail, the major supply line of the NVA and VC. This, of course, meant there were many bad guys in this area and, therefore, our safety was questionable.

One bar close to the base was where we went when we ventured out. I noted that some Air Force personnel spent time there. They were from a radar and radio site stationed there. and they became very comfortable going to this bar.

A young Vietnamese kid, maybe ten or eleven years old, hung around there. This kid would run in, act like he was throwing something at us, and then bust out laughing. At the time I saw him do this, I thought that it was a very weird and dangerous thing. His stance, his arm and body movements, and the look in his eyes made me think that he was imagining throwing a grenade or "fragging" us. I knew this kid had grown up in the war, and something in his facial expressions looked like pure hatred to me. I talked the other pilots with me into leaving because I was so unsettled by this kid's actions. As we left, I called out a warning to the two Airmen that were there, "Hey guys, we're out of here. That kid's imagination is working overtime, and he's acting like he's fragging us for real."

They laughed at us as we exited, but I heard one of them say, "You guys are beyond paranoid. Paranoid pussies."

As we left, we were looking for this kid outside but didn't see him. We got about 200 yards away from that bar, and then we saw him. That's when we heard him yell what sounded like a loud "Whooop!" Then he ran a couple of steps into the bar. After a second, he was back at the door, running out as fast as his legs could carry him.

Seven seconds later, "BOOM" the grenade went off. We tried to head back to see if we could help, but within a minute, throngs of people were running around, so we headed back to the post to write up what happened. The next day we heard that both Airmen died from their frag injuries. We did not head back downtown after that for two reasons: First, because we were warned not to by our company commander. Secondly, more than anything, we wanted to find that kid, but we knew we wouldn't have wanted to face the consequences for what we would have done if we did find him. The war was closing in as more American troop withdrawals occurred, and the NVA and VC were filling in where the Americans left. The withdrawals made the situation more dangerous by the day.

THERE IS A PRICE TO BE PAID

In any endeavor, especially fighting a war, there's always a price to be paid. The payment can be divided into different categories, and certainly not all of those require financial payment. Sometimes soldiers pay by having to adapt to artificial limbs. For some, they pay the price by undergoing multiple surgeries and adapting to the disfigurement of burns. Soldiers pay not only in physical, but also with mental tolls. Some must deal mentally with the issue of having been shot and living with PTSD for the rest of their lives. They pay by being unable to adjust after they return home. All these payments take a toll. These costs burden not only the soldier, but also the people in a soldier's life. Some of these costs are more than many soldiers and their families can handle. Any soldier who cannot put it all behind will find that the cost will wear on them and their family. Whether or not a soldier can make it will create huge stresses.

After returning home, some soldiers may discover that the

work environment is not conducive to extended issues they may be dealing with from the war. Anything can happen. Some soldiers lose their jobs and become unable to pay their bills. Families fall apart; some become homeless. All these situations can build up to the point of exploding, and that is when a real possibility occurs that the soldier decides to end their life. Certainly, such loss brings even more distress to the surviving members of their circle of family and friends.

All these issues are severe, and they destroy many soldiers as well as many families. Often the soldier cannot leave the war behind them. What they saw and did in the war is beyond the scope of human understanding of most people, and sometimes, that includes the fighter.

Costs associated with PTSD, formerly called battle fatigue, are great. In previous wars, battles might have occurred every one or two weeks. As mobility in wars increased, so too did the number of battles a soldier may be involved in increase dramatically. In Vietnam, the advent and implementation of the helicopter meant that soldiers could be transported and moved with great frequency. They could easily be engaged in multiple battles twice or sometimes three times a week. For the helicopter crews, the number of battles was exceptionally high. Depending on the number of battles in each area occurring simultaneously, the flight crew may go in and out of multiple battles in a single day. When the helicopter flew into a battle zone area, likely, enemy forces would be shooting at it. The helicopter crews knew this, and yet a long list of soldiers hoped to be trained as helicopter pilots and crew members.

Helicopters served a long list of purposes: inserting and then moving troops into the battle and moving them to fight the enemy in multiple locations, a situation making for a running battle; medevac wounded troops from the battle to advanced care locations such as a field hospitals; using armed helicopters (gunships) to aid in the fight with the helicopter taking on the most volatile areas with the mission of helping to capture land from the enemy.

Armed helicopters existed for one reason: to bring the fight to the enemy with overpowering firepower. This purpose put helicopter crews into fighting, often intense fighting, and this could and did lead to severe PTSD among helicopter pilots. I speak from experience. The difference in helicopter gunships and fixed wing fighters/bombers was the speed and altitude of the helicopter. The helicopter could be at any altitude up to about 20,000 feet and as low as treetop level or lower when not in treed terrain. It could be going from very slow speeds up to 140 kts (approximately160 mph) for the Huey gunships or up to 190 kts (approximately 218 mph) in Cobras.

In many cases flying C or M model guns in the mountains of the central highlands of Vietnam, we would start our gun runs at 60 kts. (69 mph). This gave us stability and a better view of the target and surrounding area. This proximity meant basically being in a nose-to-nose gunfight. The enemy could easily see us, and they had the possibility of hiding behind trees, rocks, or hills; however, we had a good amount of ammo. If they did not hide well, we could see them, or if they broke cover to shoot, we could take them out. Many aircraft were shot up with the crews being hit or shot down, and the crews wounded or killed, but we took a toll on the enemy that was exponentially higher than their toll on us. In a war that judged success by body count, the American fighting forces won by a very large number. The Army officially put North Vietnamese fighters' deaths at 1,100,000 vs 58,220 American fighters. All allied fighters' deaths were approximately 6,000.

What the helicopter could and did do in the fight is extensive and even beyond the ways already described. The helicopter would carry food and sometimes even more importantly, resupply ammunition. The dangerous job of resupplying ammunition took place at the forward edge of the battle area. As already mentioned, the air crews knew they would become the primary target in the battle. Even while being covered by armed helicopters (gunships), a significant risk always existed in this flight. Most crews accepted this risk because they knew that without this resupply, the friendlies would be overrun and then killed by the enemy. This stage of the battle for the flight crews was almost, if not, as dangerous in many cases as medevacking the wounded soldiers.

As a gunship pilot, I had learned that it took a stressful kind of discipline to sit there on the ground with the blades turning and the noisy engine roaring away while I waited for the troops to load the wounded. I could often see the muzzle flashes of the enemy troops shooting directly at me, but I'd never think of turning inbound, even with the fight and the enemy firepower coming right at me. The words "Above the Best" were printed over the entrance to the primary flight school entrance at Fort Wolters, Texas. My fellow pilots and I in Army Aviation took it to heart. We refused to let the best war fighters in the world be without the aviation support they deserved and needed.

When called upon to sit on the ground in a fire fight, the slicks would not leave the ground until the wounded were loaded, and then we would depart rapidly to get the wounded to advanced care for their injuries. We were not trained medevac flight crews but just the normal helicopter crews, slicks, doing our job and more. We approached our duties of inserting and extraction of troops the same way. The job needed to be done, so we did it. Flying into these hot areas was what we did, and we did it with courage every day we flew. This was just part of it: daily jobs, critical jobs as those that needed to be done by the gunships, heavy lift aircraft, and regular medevac every day, all day. The loss was high: about 2500 flight crews and quite a few passengers in about ten years speak to the danger we faced every day and in virtually every flight during our tour.

All flight crews are volunteers, and all know the danger

and the potential trauma of being shot down and captured. They all knew and understood the long hours and challenging work it takes to keep aviation assets up and doing the job. All the major categories of helicopter aviation assets—gunships, slicks, medevac, lift and heavy lift—in combat will often be in close proximity to enemy forces. It takes a certain amount of determination and belief in yourself and the rest of your crew to turn inbound and fly straight into the enemy firing at you and/or the ground forces. You must fly straight in and take the fight to directly to the enemy to relieve the pressure on the ground forces. Then hopefully, you eliminate the enemy completely. Other assets might find themselves in the same kind of prolonged fights as the gunships but performing their duties and using their fire power in a defensive role. The rest of the story also applies to them as well.

The rest of the story is all about that price to be paid. PTSD, major issues adapting to the non-war way of life, and the inability to adjust to life in the "world" add to the high cost. Part of the price is a litany of problems. There's chronic sleeplessness. Staying awake and being extremely aware meant staying alive. Now it can't be turned off; memories come flooding in and won't go away. Sleep becomes impossible. If any dreaming does take place, it's always about the harshness of battle life, the kills made, and the battle buddies killed. Sometimes it's about the effects of Agent Orange.

Agent Orange was just one of the defoliants used, thusly named because of the orange circle on the drum to identify the defoliate. Agent Orange is described as a herbicide, and it is a mixture of two herbicides 2.4.5-T and 2.4-D. Traces of Dioxin that have been found in the mixture have caused major health problems to those who were exposed. Agent Orange has been sprayed through many means including by helicopter. When using helicopters to defoliate in 1971 in 2 Corps, the spray aircraft was covered by two gunships flying above and behind the defoliating aircraft.

One memorable day we were covering a slick that was defoliating when the spray flew up and into our flight path. I could see the plumes of the spray as it wafted down over the green jungle. The air was hot, very hot, and the vivid, silvery cloud spiraling out over the dense foliage appeared like an unseasonal covering. I watched as it blew across the widening landscape, mesmerized by the undulating designs it made. I thought about a similar scene being played out by crop dusters across Kansas. But this was no wheatfield, and somewhere in my gut, I suspected that Agent Orange was more deadly than the spray going on the wheatfields. We had the front windows open, the back doors opened and pinned, and the front vents open. The moving air helped alleviate the blistering heat.

Then suddenly a blast of spray swept through our aircraft, carried on the wind. We were all coughing and spitting, cursing the damn spray. Suddenly, for safety reasons, we were at the end of the spray mission. We all continued the hacking and coughing. We flew to the airport in a daze, got out, refueled, and tried to clean the interior. The A/C from the spray aircraft walked over to us and checked on how we were doing. He was sure we blamed him and his crew. We didn't know what caused this mess, apparently a failure of some kind on some level and on somebody's watch. We calmed down as we cleaned up, but we had some serious questions.

After arriving at our base, we asked the weapons and spray experts what this poison would do to us. They assured us that it did no harm to people, only plants. However, the next day we had a joint mission with the crew from that same unit, and we saw an aircraft parked at the end of the runway with all the doors opened. We shut down and asked at the end of the mission what the deal was with that aircraft. One of the crew explained, "We're just not sure how long it will be out of service. They had a line blow out. The one that went to the container."

We noticed a clear, sticky goo was all over the inside of the aircraft. "You don't need to have any worries," he continued. This stuff's harmless to people." All they could do was take the aircraft out of service and let it air out.

We were still uneasy, so we took it up with the Commander. "Let me be perfectly clear," he said looking us up and down. "We don't anticipate any health issues because this spray is not dangerous to humans." He couldn't have been more emphatic and surer of himself.

Fast forward to today, and here we are now living with the well-documented health problems caused by this poisonous crap. I know my health issues due to Agent Orange are many and profoundly serious. Once again, my gut instinct at the time was 100% right.

So while I can try to take you through this writing in a way that makes you see and understand what is was like in the cockpit of a UH-1H (slick) and UH-1C and a UH-1M (gunships) in combat, I also want to communicate the way it was after the war. Almost to a person, any helicopter pilot I have met has said he would go again, even knowing the cost of doing this job.

One of the most wide-spread costs of the war is combat PTSD. For the soldiers in the fight, PTSD is common and serious, deadly serious. No one thing causes this syndrome, but I can speak to the situations that create PTSD, and it boils down to the pure brutality of combat. PTSD comes from the way the mind deals with the battles and bodies. It comes from seeing firsthand the effects of the Viet Cong and the NVA and the killing that is war. It comes from seeing the other side fight with no moral compass, not following the laws or limitations of the Geneva convention. This lawlessness creates the atrocities and rewards the capture and killing of non-combatants. The US and South Vietnamese prosecute offenders, but the other side, North Vietnam and its supporters China and even Russia, rewards these violators.

I am finally at peace, for the most part, but I still pay the price and always will. When your job and the very survival of you and your fellow soldiers depends on your actions, you act. You act first and later the haunting comes. When a whole group of people, both men and women, fall dead as a direct result of your shooting, the memories will have a sobering effect. If you're human, you'll be left with memories about the horror of war, and about you, your actions and yourself. Later when you are trying sleep, the thoughts and pictures in your mind come streaming back in a chaotic dance to play out the actions of that day. You may very well see the bodies fall again and feel any lack of true compassion for these people in this kill or be killed world of war. In the final analysis, you and your crew, your fellow aviators, and the grunts on the ground were all potential victims and in need of your protection. You won't know all the details, but you can assume the people shot and eradicated may very well be some of the people who had terrified innocent villagers, kidnapped teachers and the village elders, and tortured medically trained people. They could have been the ones using force to put innocent civilians into retraining camps for very long periods of time because the people of Viet Nam did not want to support Communism as a way of life. The enemy believed in force, not humanity.

Did I ever wonder if I had become something I could never return from? Yes, I do still wonder that. I still have those thoughts and pictures in my mind. It's a fact that the war changes you, or at least it did me. I know I will never be the same easygoing person that I had been before I went to war. I accept that fact.

At times, I will wake up sure that I am right back in the fight, and people are trying to kill me. I will wake up in a cold sweat truly believing that I must fight to stay alive. In my mind, the battle still rages around me. I may be at home safe in my bed with my wife, but I'm in that battle. The reality is as real and as visceral as the reality created as I sit and type these words. I experience the battle through all my senses. I hear the screams. I hear the screams of the wounded and the prayers we whispered when they were loaded. I hear bullets as they zing by me. I feel the wind of the chopper blades and see the flashes of the big guns below me. I taste the bittersweetness of the air as I take what could be my last breath. I feel the intense heat and the blazing sun. Living through these events again and again is mind-numbing, with little or no relief when you realize you've had a flashback. Once I realize that my current reality is not the war, I try to not wake my wife. I get out of bed and slowly make my way to the living room where I will sit in the dark, safe space and unwind. I will sleep no more on this night.

The logistics of simple things of daily life can also be troubling. When my wife and I go to a restaurant to have a nice dinner, she waits for me to pick a table. I must find the best place where I can be comfortable. I will pick the safest way to face, and then she will sit down opposite me. I know she will warn me if someone is coming close to me from behind. This ritual is the start of our "relaxing" meal.

I know I can go into a fit of rage when I least expect it, so I have to know how to control myself. The worst case in recent years for me was not long ago while I was riding a Harley Davidson motorcycle. Suddenly, I was no longer on the highway; I was in the air flying a gunship on a specific, memorable flight. In my mind, I was trying to control the helicopter while the only inputs I had were the motorcycle controls. I couldn't make myself snap out of it, and I was truly lost. Even as I eventually I realized I was riding a motorcycle, I found it impossible to come all the way back to the here and now. I knew I couldn't stop because I would drop the 1000-pound motorcycle on my leg, and there was a lot of traffic. I managed to keep the big bike on the road and moving forward.

When I finally found a road that would take me to my home, I was almost in what could have been a fatal collision. Had it not been for the skill of the driver of a PT Cruiser, death would have come for me. Ironically, I had survived the war, but because of PTSD, I'd brought the real possibility of death right back home with me. The winding road to my house goes through the hills of the Texas hill country. I came around a curve and could not keep the Harley in my lane, my reactions still severely slowed by the flashback. Two cars were side by side on their side of the road when I realized what was happening. My only thought was *this is where I die.* Somehow that PT Cruiser swerved and missed me by only inches. After the event was over, I realized I was almost another motorcycle fatality. The investigators would have wondered why I was in the other lane with no known reason for being in it. At the time when I had the flashback, I had been unable to stop and call to advise someone to come get me. My friends and family might have thought that I committed suicide by Harley, not caring about the possibility of taking innocent people with me. To my best recollection, this incident of the many PTSD incidents in my life is the only one that has occurred while operating a motor vehicle and putting me back in the cockpit of a gunship I couldn't fly. Since this book is about how the war looked and felt to me and not about my personal fight with PTSD, I will leave it here and to the medical experts at the VA.

LESSONS LEARNED

1. I learned how to act under fire and maintain my calm. I learned to maintain my cool, so I think more clearly. I found that staying calm allowed me to think quickly, but thoroughly. Knowing that others relied on my sound judgment and actions for their lives, allowed for no room to panic or even overreact. I had seen one pilot overreact to a complicated and dangerous situation and learned there was a better way to handle it. He was on short with a load of troops he was bringing to the battle. We were in opposition and covering him when he saw, or thought he saw, an NVA soldier with an AK 47 pointed at him. He pulled maximum power and tried to climb out, but unfortunately, he hit a big tree. The rotor blades stopped immediately, and he was about 80 feet in the air. It was almost like time stood still as the aircraft seemed to be stuck in that position. Seconds later, gravity took over, and the aircraft fell to the ground. In what seemed like time sped up, the troops and the crew

exited the crashed aircraft. Amazingly, no one appeared to be hurt. The crew ran to the aircraft behind them and jumped on as the aircraft dropped off all his troops. The troops from the crashed aircraft raced back to where they should have been, but they were told to go back to the landing zone (LZ), jump on the next aircraft dropping off troop, so they could be seen by a medic. All this chaos could have been avoided had we just made a right turn while calling the enemy to us, and we could have eliminated the threat, and all would have been okay.

2. Just a couple of days before the end of the month, I turned twenty. I felt no different, and I looked no different, but those who had a problem with a teenage gun pilot could now relax. But I doubt if they did. One more month down. The strangest part was I was feeling at home in this hell hole. Now that thought scared me.

INDIVIDUAL FLIGHT RECORD 1971: OCTOBER 1971

INDIVIDUAL FLIGHT RECORD A. J FLIGHT CERTIFICATE - ARMY

For use of this form, see AR 9544, the proponent agency is Office of the Assistant Chief of Staff for Force Development

28. LAST NAME - FIRST NAME - MIDDLE INITIAL

McCARTHY, JOHN M.

1971, OCT

29. SERVICE NUMBER SSAN — 545-86-5237

30. GRADE AND COMPONENT — WO 1 USAR

SECTION V - FLIGHT HOURS ACCRUED - TOTAL HOURS FLOWN BY MONTH *(For Local Use as Desired)*

JULY	AUGUST	SEPTEMBER	OCTOBER	NOVEMBER	DECEMBER	JANUARY	FEBRUARY	MARCH	APRIL	MAY	JUNE

SECTION VI - RECORD OF FLYING TIME

			FIRST PILOT FLYING TIME								COPILOT FLYING TIME						
			FIXED WING				ROTARY WING				FIXED WING				ROTARY WING		
					NIGHT				NIGHT					NIGHT			NIGHT
DATE	AIRCRAFT TYPE MODEL SERIES	MISSION SYMBOL	AIRCRAFT COMMANDER	INSTRUCTOR PILOT	FIRST PILOT	WX INST	VFR	WX INST	HOOD	WX INST	VFR	WX INST	HOOD	CO PILOT	WX INST	VFR	WX INST	WX INST	VFR	WX INST
a	b	c	d	e	f	g	h	i	j	k	l	m	n	o	p	q	r	s	t	u
OCT																				
1	UH-1M	C			2.3															
2	UH-1M	C			1.5									0.5						
3	UH-1M	C	1.5																	
4	UH-1M	C	2.0																	
5	UH-1M	C	3.1								0.3									
7	UH-1M	C	3.1																	
9	UH-1M	C	2.0																	
11	UH-1M	C	4.3																	
12	UH-1M	C	5.5																	
14	UH-1M	C	5.3																	
17	UH-1M	C	1.0																	
18	UH-1M	C	2.3																	
19	UH-1M	C	1.8																	
20	UH-1M	C	0.5																	
21	UH-1M	C	7.5																	
23	UH-1M	C	1.5								0.5									
25	UH-1M	C	1.3																	
27	UH-1M	C			1.5															
29	UH-1M	C	3.8																	
					LAST ENTRY															

31. TOTALS THIS SHEET: 47 | 5 | | | | | 1 | | | | 1

32. TOTAL BROUGHT FORWARD FROM SHEET NO. 3-3: 14 | 261 | | | | | 11 | 1 | 5 | 40

33. TOTALS TO DATE: 61 | 266 | | | | | 12 | 1 | 5 | 41

DA FORM 759.1 *(PART II)*

THINGS WERE NOT ALL GOOD ALL THE TIME

The 92nd was a company that was well set up and running as far as I could tell. However, that doesn't mean stories with opposite accounts weren't out there. We often talked to other combat pilots when we had to refuel and rearm or between lifts. They were always a good source to learn from and hear what was happening. I paid close attention to what our enlisted men had to say since a random conversation with one of them could save your life. It was through this pipeline that I had begun to hear several stories of aircraft sabotage. Fortunately, we had not had any such unimaginable event occur.

Then it happened. I was at the flight line getting ready to pre-flight my aircraft when I realized that my aircraft was missing. Thinking that situation was somewhat unusual, I walked over to maintenance and to inquire. It didn't take long to locate a maintenance person, and I didn't waste any time quizzing him.

"Hey, my aircraft's missing! I told him, just shy of a shout.

"Well, I hate to tell you this," he stated deliberately, "But the crew chief found something wrong. We think it can be fixed overnight."

With nothing that could be done, I went on my way and returned the next morning. "I'm back," I announced.

The maintenance man said, "Well, it's still not ready. Once they got into it, they found another issue. A small one, but it had to be fixed. They worked on it all night, but it's been slow going, and no way it'll be ready in time for this morning's mission." About that time the Maintenance Officer appeared.

"Let me see what I can do," he said. Within a few minutes, he worked with Operations to find me an aircraft not originally scheduled to fly. It was available, so for this flight, it could be assigned to me.

I was relieved, but cautious. Since the aircraft was not my aircraft, I had the crew chief and the pilot doing a very thorough pre-flight. I had no clue if something could be wrong, but I was used to *my* aircraft. So, in the interest of safety and peace of mind, I felt that erring on the side of caution was the appropriate thing to do. Operations touched base with the ground unit and a slick waiting on us to explain the delay and that we would be taking off very soon.

As the three of us were preflighting, I chose to preflight the engine and transmission compartments. On a gunship, the rocket tubes and miniguns are attached to hard mounts. Also, a rod from the outside of the engine compartment to the hard mount helped the hard mount stay in place and not oscillate in the wind. Now this rod proved to be an excellent hand hold to climb up to the engine and transmission compartments. I grabbed it and stepped up to the top of the hard mount. I then grabbed the aft part of the transmission cowling and pulled myself up. However, when I pulled on the upper hand hold and lifted my foot for the foot hold and pulled, the cowling came off with a big bang as it landed on the pavement. I fell off the hard mount right along with it. As the aft cowling hit the ground, it then hit me. The others came around the aircraft when they heard the noise. I was knocked goofy for a couple of minutes, but when things cleared up, I was able to tell them what had happened. By the time I had pulled myself up, checked elbows and knees, and dusted myself off, my crew chief had already downed the aircraft, so it could be inspected.

In the meantime, while all of this was happening, my aircraft's repairs had been completed. It had already been test flown and returned to service. Since it was ready, I happily returned to my aircraft. We preflighted thoroughly and then off we went.

On the way to the staging area, we talked about what happened. "Chief," I said, "That kinda shook me up. I didn't know where I was for a minute. I can't believe maintenance was so sloppy. What do you think happened? They just got so busy they didn't secure that cowling? All it takes is just a few screws." I was rubbing my sore elbow as I made the observation.

The crew chief stopped right where he was, and then he looked at me and said very slowly, "Maintenance didn't have anything to do with it. I think that aircraft was sabotaged."

I put that thought out of my mind since I had a job to do. We wouldn't know for several hours what happened anyway. We went to work. It was a two-hour flight, and we did a lot of shooting under heavy enemy firing, which we happily returned to them. Despite the earlier distraction, our fire power proved to be highly effective. Between the other gunship and us, we got a fairly large number of the NVC. Then our infantry went in and finished the job.

A squad leader, a lieutenant, called to tell us that our fire had been extremely accurate. The enemy had taken away the substantial number of dead and the wounded, leaving a blood trail to what was apparently a tunnel opening. Upon investigation, a large stash of weapons was recovered. We landed to see the stacks of weapons stacked by the tunnel entrance. The squad leader gave us our choice of weapon to keep. I took a folding stock AK-47. It was a perfect size to keep beside the door and good insurance to have if we got shot down and needed to escape. I figured with an AK-47 any enemy we ran across would have ammo. With an American weapon there was no way to get extra ammo if you are on the run anywhere in enemy territory.

As soon as we were ready, we cranked up and departed for home base. With the success of the battle and choosing my AK-47, I had not been thinking about the possibility of sabotage and how close I'd come to flying that aircraft. Now that we returned to the conversation to that topic, the more we talked about it, the more I also began to think it could very well be sabotage. Sure enough, when we returned to our base, we were met by a staff guy and told we had to report to the Company Commander.

We were called in one-by one. Then we each exited through a back door, so we didn't have access to anyone else being questioned. I was the last one called in. The major asked me to sit down. He remained standing.

"McCarthy," he said, "I'm here to tell you why you're here. That aircraft you were supposed to fly was sabotaged. No question about it. Upon further investigation after that cowling fell off, we discovered rags shoved in the transmission. Whoever did this was very well versed in how to do what he did. This would've been a near-perfect murder of you and the crew. Any ideas who might have done it?"

I was mostly speechless, but answered with a deliberate, "No, Sir."

The saboteur had opened the cowling and loosened all the bolts on the top of the transmission. He was able to pry it up enough to shove the rags in. He had then tightened down the bolts and repainted the slippage lines on the bolts. Then he closed everything back up. The one thing he failed to do was to reinstall the bolts on the cowling cover, so when I pulled on it there was nothing to stop it.

I went on my way after I signed my statement. Soon after, I saw the pilot assigned to that aircraft. He had transferred to our company because he couldn't deal with the lax way his company was run. Some liked it that way, but he was not one of them. Later he told us what had happened. The crew chief for that aircraft had shown up under the influence, and the pilot had confronted him, figuring it was marijuana. He told the crew chief that he would take a day to consider what to do. The perpetrator knew that this no-nonsense pilot would definitely turn him in. Of course, he also knew he would fail a drug test. He hatched up the scheme to sabotage the aircraft in an attempt to remove the pilot, so he could not report him. Oddly enough, he would have come to me as the drug and alcohol abuse officer for his toxic screen. He knew I could not be bribed, but he did not know I would be on that aircraft on that day. In the end he confessed and was charged with four counts of attempted murder. I heard he agreed to a plea deal, and off he went to rehab. To my way of thinking, he didn't receive enough punishment for outright sabotage, especially sabotage meant to murder me and my crew.

RANGER EXTRACTION

As it is with so many of the military special teams, the likes of Green Berets, Army Rangers, Navy Seals, and many more are mostly not well known because of the type of work they do. They work behind enemy lines most of the time. When thinking back to World War Two, these elite special teams performed what many of us would say were super-human acts as a matter of course. Like the US Army normal troops, when not on a mission they train. They train to the point that normal people could not begin to do the jobs they do. Like Army Aviation, it is solely a volunteer job. It takes the best of the best to live the lives of all the special operations teams.

During the D-Day invasions many of the soldiers coming off the Landing Ship, Tanks (LST) were trapped on Omaha beach. It was a few Army Rangers that began scaling the rock cliffs with basically nothing but their highly trained minds and bodies to climb up these rock cliffs while the Germans were picking them off as they climbed. As one was shot and fell, another took his place. Of the many amazing acts of heroism in the different branches of the military at war, few, if any, have eclipsed the heroism and physical punishment of the deeds of these Army Rangers. Truth be known, it may not be common knowledge that the Rangers' heroic actions allowed the Allies to keep moving ahead. The invasion forces on Omaha beach would probably have been slaughtered by the Germans, and without the heroic action of the Rangers, the war would have been extended and more lives lost. The Rangers come from this historical tradition, and the soldiers of today are every bit as brave and well-trained as those from their past. It takes a different type of person to even consider joining the Rangers and be prepared to handle the jobs they perform in unsung glory.

I recently had a conversation with my father after I had been doing some soul-searching. "What do you think," I asked him, "drove me to want to fly directly into enemy weapons?" After a minute of silence, I could tell he was thinking about the nature of my question. I continued, "Basically the situation formed a face-to-face fight. Occasionally these weapons were radar controlled. If they weren't radar controlled, then NVA or VC operated semi-automatic weapons, using just their eyes and ears. You know, up close and personal."

I had been asked this same question, and I had no idea how to answer it. My mindset was that I had a job to do, and I did it. Unfortunately, that fact did not answer the question.

Deliberately and slowly, my dad turned the tables and asked me a hard question, "What were you thinking of when you were actually in the middle of doing that attack helicopter pilot job?"

"I don't know. I just thought it was something I could do," I repeated. Again, that did not answer the question, but it led me to the answer. Suddenly, it hit me. "I believe this was the best service I could give to the ground troops and other aviators."

It was something I believed. I really could help the troops win the battles. Essentially, it was the best way I could help keep them in a position to win against the Communists.

These Army Rangers wore that Ranger tab, and it showed what American military soldiers are capable of. This mission was the deadliest situation that could have occurred. If these troops lose their invisibility, it's like they have the plague. When this happens, there may be only one of two choices: fight it out wherever they are when discovered or start running and get out a call for help.

This one young lieutenant knew that two gunships were just across the trail, and if we were available, we could handle the enemy. It appeared to me that these troops, who had fought so bravely, were not going to be lifted out. The slick behind us was too far back to be any help, so I made a plan. Then I did some convincing of the flight commander. After a brief time, he approved it. This involved what a Huey gunship never did, which was to land right into enemy ground fire.

I saw this soldier that was in the front of these troops to pick up what was seconds away. Initially the Local Transfer Director (LTD) just stared at us and then looked to return to the fight. They were outgunned, outmanned, and fighting against troops that were rested and refreshed. We were sending radio calls and waving the trooper to get over to us when he finally understood that he was the last man standing. He moved as quickly as he could towards us as I cleared the crew chief to open fire into any enemy troop, with no possibility of hitting Friendlies. We sent a burst of mini-gun more or less at the enemy troops that were insight away from where our troops were lying on the ground after being shot.

We were about a minute and a half out on the ground, and Lead was calling on us to leave after a lot of enemy soldiers were running towards us. The crew chief got onboard and grabbed the squad leader and was pulling on him to get him solidly into the aircraft. We flew as if we were heading back to Vietnam and our quarters, but they did not know we were coming back for one run at each of them. When ready from about five miles away from them we went inbound at them. Lead went first and flew a quick inbound run firing fifteen parries of rockets at their location. I went inbound at a slower pace and let the co-pilot weapons man pick out his beliefs on where they were. I put six pairs of rockets just outside of Leads and my go-to co-pilot fire. I saw many bodies on the ground and many enemy troops running, and I aimed my rockets where they should make it to if I had not fired. Bad choice for them and an excellent choice for our men. I was taking my last look at the carnage on the ground, and then some idiot shot at us. I saw the flash of his weapon, and then he was gone as my co-pilot showed no mercy to them.

We made it back to Banh Me Thout in a short amount of time, and after shutdown, the young lieutenant threw his gear into a jeep that was taking him in for a debriefing. He looked me directly in the eyes and said only one word, "Thanks." I could feel his gratitude and hear it in his voice. We discussed the details with the other crew, and we all agreed that we had done the best we could under the circumstances.

THE FLIGHT FROM HELL

We were tasked to go as gunship protection for a mission that involved aircraft and crews from several companies in our operational area. This was almost as large as operations earlier in the war. In our time there, we did not have many large insertions and extractions. This mission was clouded in secrecy from the start. We headed to the area where we would stage, which wasn't a great distance from our home area. We arrived just at first light. Aircraft and crews, along with a substantial number of ground forces, were already in their respective areas. It all looked like a well-organized operation brewing but looks are often deceiving.

A voice on the radio instructed us where to go to a specific part of the staging area. Once we were shut down and out of the aircraft, we got together and looked at it all over. Another couple of gunships were there. We had worked with them on another operation a few weeks prior, and we had a different philosophy than they had. Immediately, those tiny hairs on the back of my neck stood up. It was almost as though we had walked into a cool area out of the oppressive heat. We had a quick conversation with everyone involved, the pilots and crews, about how we had to handle the job this day. We headed down a small hill we were working from to the briefing area.

It looked like a scene from a movie. Armed Military Police kept large briefing boards covered and guarded from prying eyes. It was all surreal. Just after we arrived, a major general stepped up and called the briefing to order. As time went on during the very thorough briefing, our crews became increasingly concerned about what we were going into. In a quick version of events, it came down to this: there were two sets of guns, and we would prep two different areas. The idea behind this was that the enemy already knew what we were going to do, and the generals had plans to make the enemy look foolish and unprepared.

This would prove to be a very deadly and mistaken idea. In the briefing, a South Vietnamese General told the Americans that a South Vietnamese spy had filtered into the enemy operations and informed our people that the enemy already had seen our plans with their own spy. Our generals devised a plan to prep the landing zone's (LZ's) original location as if they were continuing with the battle plan. Meanwhile, they had selected *new* landing zones that would bring the soldiers into the operational area with a surprise attack from a different direction.

We were designated as the attack team prepping the LZ's the operation would be using although these were now *fake* LZ's.

We left the briefing to go back to our aircraft and be ready to go in fifteen minutes. We all thought the same thing. If they knew they had a spy imbedded into this Battalion's operations, why would this operation go forward? In the world of Vietnamese planning, this operation was enough to make you a little queasy. We were getting ready to go prep the fake LZ's while the other guns were going to prep the real LZ's. We knew the enemy had our complete battle plan. We knew there were particularly good odds that we were going to meet some real resistance. No one thought this idea was very smart; however, we were told what to do, and we knew how to do it.

Off we went to waste a lot of ammunition. We treated our part of the operation as if it were real. We prepped an area for the inbound slicks with a very thorough covering of their planned LZ and lit up the area, just in case we were wrong, and the enemy was present. We prepped these first coverings of the fake LZ's and met no resistance. We were now positive the enemy knew the plan. We finished and headed back to the staging area. Radio discipline was necessary, and anyone not keeping their mouth shut would be met with severe penalties. We were listening to the action from the other aircraft, real action and real troops. We heard the first radio calls from the lead A/C. He suddenly started screaming into the radio for all the aircraft to turn around and leave the area. We could hear the rounds hitting the aircraft when he called. The last call he made was a go-around to depart the area. He then, in a very weak voice, let everyone know he was returning to the prep area, and that they were all dead. We did not know if "they were all dead" meant the enemy killed in the prep by the other guns, or some other group killed in the LZ. He made no new radio calls.

We arrived at the prep area and made the turn to the parking area we had used earlier. We saw the aircraft sitting right where it had been, and it looked odd. We moved into our designated spot and let the engine cool down for two minutes. After shutdown and no post flight damage found, we got together to go look at the aircraft that had been shot up. We saw what appeared to be transmission oil pouring out of the drains on the bottom of the aircraft and realized that blood was pouring out, as well. As we moved in closer, we could smell the blood and could see what was left of the windshield. We approached the left side door and could see that it was riddled by their AK 47's. No crew members sat where they should be. Once we approached, we could see the bodies still on the floor. No one was alive in the aircraft. It turns out that they had a group of eight Koreans who were all dead on the floor, as was the crew chief and gunner. The co-pilot and pilot were gone, also. Twelve bodies.

As things turned out, the pilot wasn't dead. As the only survivor, he had a questionable chance of recovery. He had shown unbelievable skill as a pilot. We heard about some of his injuries from a Captain from the briefing team who filled us in on the unbelievable injuries he had. A truly brave and tough man flew that aircraft.

It turns out the NVA had used a trick we had found the VC using. They would lie down in the grass and weeds while another soldier covered them up, so they couldn't be seen. The guns that had prepped the LZ had not shot into the ground on the approach path. We had a complete inquiry into why they didn't prepare the approach path. We had talked to them, and they believed, very strongly, that prepping the ground would be the optimal way to do things, but doing so would give away the approach path. We had seen nothing to support their view.

We believed that the VC or NVA would lie down on the approach path in wait, as we had seen that day. It was a serious difference of opinion, and we had been shown that we had been correct in our view of prepping our LZ's. It was a case of being dead right, but people were still dead. We would rather have been wrong and for them to still be alive.

This was the bloodiest aircraft I had ever seen. We did not count bullet holes because we had to get out of the way of a team going over this aircraft. There would be nightmares over this in the long run and for a long time.

The operation was terminated while they tried to find a spy in their mist.

SHOOTING MONKEYS

In every job, I believe you'll encounter some tasks you might not want to do. Perhaps the task is difficult, or it may be messy, or it may go against some of your core feelings and beliefs. In any of the cases, it's very difficult to make yourself do such a job. I came upon this very situation one hot afternoon in early September.

We were called out to assist some troops who were trying to move a lot of monkeys off this hill. Yes, monkeys. The terrain where the monkeys resided proved to be a bit of an anomaly in our area. The South China Sea was just a few miles east of our location, and the mountains began just about fifteen miles to the west of us. However, just about six or seven miles to the northwest of our base was a lonely, large hilltop. From that high point, the VC and NVA would mortar our base. In return, our artillery would fire 105mm shells out of M101A1 howitzers. Supposedly, this situation led to VC or NVA troops picking up whatever unexploded rounds were fired but did not explode.

So, it was thought that we had a kind of revolving door on the ammunition while we were firing to kill as many VC or NVA troops as we could. I cannot say how many unexploded rounds were there, but in my thinking, it seemed highly unlikely very many failed to explode. I believe, even to this day, that someone decided it didn't look good to have gunships and ground troops hunting down monkeys; however, these monkeys were causing serious injuries to the ground forces, so removing them was actually a correct move that needed immediate attention and action.

This hot afternoon we were sent out to cover the grunts going up the hill. We were sent to the location for an unspecified reason, so we were curious what we would be attacking on this day. Soon we saw the monkeys. We were getting into position to shoot when we had our lead call back to double check that the monkeys were our targets. These were not your basic Rhesus monkeys; they were large gorilla-looking monkeys. Our communication provided us with some valuable information. These were not gorillas, but they were a smaller species that looked and acted like the true gorillas. He was told that these monkeys were jumping from some high spots in the trees to land on top of the troops. This was no joke; two infantry troops were severely injured by these attacking animals.

As we started circling this solo hill-top, we devised how we would attack this problem. I was moving slowly in a counterclockwise pattern when I saw one of these apes screeching and howling at us as he hung on to the tree limb. I picked up my M-16 and pointed it out of the window directly at him. He was pounding his chest and making noise when I decided I had to shoot this mean bastard. As he continued to make threatening noises and gestures at us, I shot at him. To my surprise he grabbed his chest and squealed a blood curdling yell as he fell to the ground. All at once, I felt bad because this monkey had no political affiliation. He wasn't fighting to force

Communism on people who didn't want it. These monkeys were part of the Viet Nam landscape, and they were attacking from the trees the way they'd attacked since the beginning of time. They didn't recognize uniforms or political leanings. By the time we arrived, they were conditioned not to be afraid of helicopters, and they had proven to be a serious issue for the troops.

I did not like that I had to shoot such an animal in his own jungle environment, but I also felt bad for any troops that may have encountered him. Certainly, this was a different day than most of our days. We spent the afternoon clearing out the trees, so the monkey habitat was gone, and the monkeys could more easily be completely removed. Finally, we did have to go back in after refueling to run the last of them out. There were times to fight and times to stand down, and on occasion complete some very strange missions.

Life works in strange ways. I wondered why shooting monkeys bothered me and shooting the enemy did not. For the most part, I know the obvious answers: following orders, strong belief that what the enemy was doing was horribly wrong, defense of our ground forces and those of our allies etc. However, I believe it comes down to armed vs unarmed.

These monkeys never had a chance once we went after them. All of that explains how this was done and the fact that if we had not intervened these very same innocent monkeys would still be attacking our ground forces, including our allies. The lesson is: look at other's situation before deciding what is right and what is wrong. I knew this, but I had forgotten it, and what I saw were innocent monkeys, but what the ground troops saw were animals that were very much stronger than humans and trying to harm or kill them. The reality was simple. Remove this actual threat and protect the ground forces. We were doing our job, and it just happens that the threat was not human.

FRIENDLY FIRE

Of all the things heard over the radio, one message will turn your blood ice-cold instantaneously. That call is "Cease fire! Taking friendly fire!" It gets repeated with more volume by the next call and more volume with the next call and the next and the next—until it stops. The demands of a gun pilot are extensive and the level of importance always increasing. The guns will be watching every ship in the flight and monitoring multiple radios. Often the fight increases with both sides because winning isn't just winning, it means survival in most cases. When you hear "Taking fire! Taking hits! Going down!" that call changes everything from what patterns you fly, to changing where you need to concentrate your firing, to calling back for Medevac. When you hear "Friendly Fire," all shooting by Friendlies stops. In the initial call, you don't hear where the firing is coming from.

As a gun pilot, I was putting out a lot of ammo, and I when I heard that awful call, I knew instantly that a strong possibility exists that one of us may have just killed an American or South Vietnamese soldier, the very people we were there to protect. I may have done it.

We had an incident one night, and we all almost became physically ill. A night patrol had been caught in a trap and was getting lit up and starting to lose ground. The platoon leader, a captain in this case, immediately called for gun support, and it went to our operations. We had crews standing by, so they cranked up in about two minutes. The captains, two aircraft commanders, had just joined our unit about two weeks prior. They were both second-tour gun pilots who knew each other from their first tour. Our first impression: they were good pilots, ("sticks") with level heads. They were good guys.

On this evening, they were on standby, which gave two of our crews a break. So, it was a relaxing night, as much as you relax when you know the enemy will rocket you, mortar you, or try to slip in through the mines and stay out of sight of the guard towers. This evening, we had been playing cards, but I knew I when heard the engines winding up, they had bounced, so we all headed up to operations to see what was going on. These two pilots had shown they could do this job, so no one seemed apprehensive about the two new guys covering this. They had done all this before in their first tours, so we were just being nosy.

They got out to the battlefield quickly and started taking fire immediately. We wondered if the enemy knew they were gunships because shooting early like that just told the guns where at least some of the enemies were. They quickly got the attack going, and it sounded like it was going well. Suddenly there was the call: "Cease-fire! Cease-fire! You are shooting friendlies!" We were all heading to our aircraft but were stopped immediately. We were not wanted by the grunts. It turns out the friendly fire had killed two of our troopers. In what would be light speed today, another team of guns were on the way from across the street. They arrived quickly, and our team was sent home—in shame.

Once they arrived, we went out to meet them, but they were kept from us by some military police (MPs) who appeared out of nowhere. The crews were moved to a room in operations, and we were sent to our rooms. It was then that our missions for the next day were canceled and given to some other unit. We were told to stay out of the way because a major investigation, potentially a criminal investigation, would start first thing in the AM. A lot of activity ensued all night long as aircraft flew in, and we heard people in jeeps talking, but we were kept totally in the dark. We literally received no information whatsoever. That brief encounter with a disembodied voice over the radio was all we knew. Period.

The next morning, we all went to find food and were forced to eat in the mess hall, much to the surprise of the cooks. The food was fairly good, much to our surprise. We tried to sit around and follow what was going on without much success. As if we were being herded, the Commander sent us to the Officers Club, which had to be opened for us. In direct terms, we were told to stay out of the way. Then the Sergeant Major came in to speak to us. He showed compassion and understanding for us as we all tried to work this out, but he could only tell us one thing: "Men, the investigation has started. Be ready because they're pulling all your records. I repeat. They're pulling all gun pilots' records for audit. They're looking for any irregularities."

A reaction of disbelief spread across us. We could not understand why they were pulling all our records.

"This is a necessary step," explained the Sergeant Major. "Even the Major and the Platoon Leader are being checked out. It needs to be determined whether or not this group has a problem."

First we couldn't believe our ears. We stared at him in

disbelief. We went from being offended to full-blown angry. It was a witch hunt. Since we were the only ones in this part of the mess hall, we had the opportunity to talk with each other and make a unanimous decision. We would not play their game. We would answer legitimate questions but no leading questions that would be disparaging to our Commander and his staff and, most importantly, to our pilots.

We knew they could not do anything to us for being part of a coverup because we didn't have enough information for a cover up. If the investigator started asking leading questions, we would answer by saying we were turning in our wings. As things progressed, it became clear that their questions were designed to "back stab" serious soldiers. We could not function as pilots under these conditions. We knew we were within our rights, so we weren't worried, but if they decided to take us up on our threat, we would turn in our wings right then.

We all put a pair of our wings in a pocket, so we could hand them over to the investigator right then. It was drastic, but we're part of a team in a war where they ask so much of us, and now it seemed as if they wanted to punish us. My thought was: good luck replacing all of us. We knew if they did take our wings, heads would roll. The removal of qualified and experienced gun pilots with no replacements available was unthinkable.

The ridiculous part of all of this was we only knew what we had heard over the radio. The platoon leader on the ground did not know who killed the troops in this firefight; he just jumped on what he thought and reacted. In other words, he had no definitive information. At the end of the day, we were told our company was "on hold" until an autopsy was done on these bodies in the United States.

I was called in for questioning by the Commanding General the next day as were some of the others. We were interrogated in selected and secret ways, all designed to make us say one by one by one that the pilots had not followed our operations and training rules. It was as if they didn't know we were just listening to the radio transmission. As it was, we did not hear anything unusual beyond the haunting words, "Cease fire! Cease fire! You are shooting friendlies." They did not want to hear us report that we knew nothing.They were looking for some sinister plot that simply didn't exist.

My thoughts were to try to expose the Colonel who decided to accuse us, something the grunt Lieutenant wanted. I was selected to be questioned, and I guess being the youngest played into the decision to grill me. It was a terrible and confusing time. Those of us who *weren't* selected for questioning didn't understand why. Those of us who *were* selected didn't understand why either.

When it was my turn, it went as I thought it would, of course. The first question I remember directly was, "Was this a deliberate act?"

What? Did I hear the Commanding General right?

He repeated, "Was this a deliberate act?"

I was appalled at this question and took a deep breath before I answered. I wanted him to think he had scared me when, in fact, he just made me furious. I answered by saying as emphatically as my anger would allow, "Honestly, I do not know what you are talking about. You've kept all of us in the dark since this incident happened. I don't know anything. I don't know what happened. I don't know why it happened. So no," I continued, "I know nothing." I put my hand in my pocket and felt the solid metal of my wings. I prayed a little at that point.

The Commanding General took a long, burning look at me. When he next spoke, I could hear the seething anger in his voice, "You, soldier are a smart ass. I asked you a direct question. If you fail to answer, you're going to have a huge problem."

I took another deep breath and clenched my jaw. I paused deliberately for a minute, looked him straight in the eye and said slowly, "I. Don't. Know Anything." I had nothing more to say.

That set him off and he screamed, "You better have something to say, you smart ass."

I reached for my wings and set them on the table. "I'm resigning my wings and flight status. I request that I be returned to the States before you ask any more questions of me. Furthermore, I want an attorney to be assigned to me."

The Commanding General looked at the court reporter who was keeping track of every conversation. "You," he yelled. "Stop with that." He obviously didn't want the intimidating tirade that was about to ensue recorded for history.

"You idiot! Talk. Don't you know you could go to jail and spend time in prison for these murders," he screamed. For a moment, I thought he didn't hear what I'd said. He didn't acknowledge my wings on the table.

I just sat there and took it, looking at some point in the distance. I knew he could continue to ask the questions, but I could continue not to answer. I looked at the court reporter and said, "Please keep track of this conversation. I have a right to have a record of all of this." The court reporter turned the recorder back on.

Then the General strode over to the recorder and slapped it off. The court reporter looked perplexed. The General opened the door and growled "Dismissed." I grabbed my wings on the way out.

When we all got together to compare notes, almost the same things had happened to every other pilot, so I knew we were in no real trouble.

It was two more days before the lab results from the autopsies came back, but the metal in their bodies was not metal from American weapons. It was from hand grenades made in China. Suddenly, this crazy, monumental situation went away as if it had never happened.

Another event in my short notes from Vietnam. I was a hardcore career soldier, or so I thought; however, these events were so horrendous. Those in command placed us in a position where they hoped we'd turn on each other by scaring us to fall for their flawed logic. This was the first time that I thought I might *not* make the military a career.

LESSONS LEARNED

1. I learned not to become so bogged down in the idea of how a mission should be conducted.

2. I learned that everything I planned on could change, and often will change, at the worst time possible.

3. It may be necessary for me to adapt and overcome something, and I may have only mere seconds to act.

4. I know that I'm putting myself in grave personal danger with any decision, but I have to act on faith that I'm doing my best. I may be able to save only one soldier, but that makes my risk worth it.

INDIVIDUAL FLIGHT RECORD 1971: NOVEMBER 1971

INDIVIDUAL FLIGHT RECORD AND FLIGHT CERTIFICATE - ARMY (PART II)	26. PERIOD COVERED 1971: Nov-Dec	27. SHEET NUMBER 3-5
28. LAST NAME - FIRST NAME - MIDDLE INITIAL McCARTHY, JOHN M.	29. SERVICE NUMBER SSAN	30. GRADE AND COMPONENT WO1 USAR

SECTION V - FLIGHT HOURS ACCRUED - TOTAL HOURS FLOWN BY MONTH (For Local Use as Desired)

JULY	AUGUST	SEPTEMBER	OCTOBER	NOVEMBER	DECEMBER	JANUARY	FEBRUARY	MARCH	APRIL	MAY	JUNE

SECTION VI - RECORD OF FLYING TIME

			FIRST PILOT FLYING TIME										COPILOT FLYING TIME							
					FIXED WING				ROTARY WING					FIXED WING			ROTARY WING			
						NIGHT				NIGHT					NIGHT			NIGHT		
DATE	AIRCRAFT TYPE MODEL SERIES	MISSION SYMBOL	AIRCRAFT COMMANDER	INSTRUCTOR PILOT	FIRST PILOT	WX INST	VFR	WX INST	HOOD	WX INST	VFR	WX INST	HOOD	COPILOT	WX INST	VFR	WX INST	WX INST	VFR	WX INST	CROSS COUNTRY
a	b	c	d	e	f	g	h	i	j	k	l	m	n	o	p	q	r	s	t	u	v
Nov																					
1	UH-1K	C	2.5																		X
2	UH-1M	C	1.5																		X
3	UH-1M	C	1.3																		X
4	UH-1M	C			2.7																X
5	UH-1M	C	3.3								0.3										X
6	UH-1M	C	1.0																		X
7	UH-1M	C	2.5																		X
7	UH-1M	C	0.5																		
9	UH-1M	C	0.8								0.5										N
10	UH-1M	C	2.3																		X
13	UH-1M	C	1.0																		
18	UH-1M	C	0.1								1.0										N
20	UH-1M	C	5.3																		X
21	UH-1M	C	4.4																		X
26	UH-1M	C	1.5																		
29	UH-1M	C	1.8																		
										LAST ENTRY											
NO TIME FLOWN DECEMBER																					
31. TOTALS THIS SHEET			40	3							2										
32. TOTAL BROUGHT FORWARD FROM SHEET NO 3-4			61	266							12	1	5	41							
33. TOTALS TO DATE			101	269							14	1	5	41							

WINDING DOWN

After the long Monsoon rains finally came (they had lasted close to an interminable two months), we flew less. We stayed close to the Ho Chi Minh trail because the big buildup by the NVA was obvious. Now and then they screwed up, and that gave us a quick look into what was happening.

Sometime in November, we were returning from a mission along the trail when suddenly something unusual caught my eye. In the moment, it was almost as though a vision suddenly appeared. I thought to myself with what felt like some clarity and insight: *This seems extremely important.* From an outside point of view, that thought might've seemed incongruous because I was staring at a big clump of red clay. Although it didn't seem to fit in, there it was in plain sight. Suddenly, it moved, and my premonition met reality as I realized we were in a bit of trouble. As the vision came to life, I knew that the movement I saw was a tank's turret. At that very moment, it was turning to get a firing solution directly at us.

I knew at once that I had to move in the opposite direction from this turret, and I had to move NOW. Once I turned, I quickly fired a 2.75-inch Folding Fin Aerial rocket at it.

Success! I hit the tank, but other than knocking some red clay off it, my rocket barely fazed the big mass of red mud and machine. We wasted no time. We knew we had to leave since we had no information on this NVA operation. We knew the rockets we fired would ultimately not help our cause at all. As we moved out of the tank's way, we knew that Military Intelligence (MI) needed to be advised of this turn of events. So be it from vision to reality.

This would just about end my tour with the 92nd Assault Helicopter Company. I was moved to B Troop 7/17 Calvary, which was still highly active in the war while the 92nd would be closing up and returning to the USA.

I thoroughly enjoyed my time in the 92nd and flew with some excellent pilots and crew members. While no one single soldier here, or the company, became famous, it's safe to say that we just did our very tough jobs in a difficult environment; moreover, we did them successfully. I have some great memories, and I am proud to have served in combat with this outfit.

In a moment of reflection, I have to say the following: we had some very tough missions that we did with no fanfare. We lost some crew to war's horror. Some exceptionally good Army pilots and enlisted crew members died in this misunderstood war. Others were wounded and returned home in worse shape than they arrived in. For most of us, we were there to fight the spread of Communism and preserve the right to self-determination. We all knew the risks and accepted them willingly, not to be confused that we *wanted* to die in this country, but that was the risk.

During this period, I saw the units standing down and returning home. As we continued to fight in this war, it was with an ever-decreasing level of aircraft and men. An eye-opening fact about this time is that the military resources were substantially decreasing almost weekly; yet the missions kept coming in. A couple of years earlier, substantially more aircraft and troops were fighting hard. Now, we were left without the backup that had been the hallmark of this war.

We were keenly aware of the huge buildup of North Vietnamese and their allies for the final part of this war that would take place to overrun South Vietnam and establish one government for Vietnam, ending the South and North division, and having one Communist government. This, in fact, did happen; however, the American influence had been strong enough that within a few decades Vietnam would turn into a market economy, and it is now seeing a huge growth in tourism, many tourists being American.

Before leaving and heading up to a Cavalry unit, I saw the problems of being the last fighters. Quite a few Aviation units were still out there functioning, but the number leaving seemed to be an ever-increasing event. As each company stood down in our area of operation (AO), fewer and fewer resources became available to help when things turned bad. When I first arrived, the availability of so many assets provided a level of confidence, confidence needed should you be shot down. Now the circumstances made the job harder mentally, but the soldiers kept fighting until their unit was sent home.

The enemy became increasingly brazen when they saw they had an even larger number of troops vs. the US troops. I saw this continue until I went home, and very few aviation assets were available. It became a different war where essentially, we were the hunted, and they were the hunters. By that I mean they would be more likely to take on an American unit knowing we did not have the reserves that we have had through most of our time in this war. All the being said when the last combat troops left in 1973, the country was still in the hands of the

South Vietnamese. The country fell in 1975, so I surely am sick of hearing that *we* lost the war. The South Vietnamese lost the war and their country's sovereignty.

SYMBIOTIC TEAMS

I t is November of 1971. I have been in Vietnam for eight months, but it feels like I have been here for a long time. I do not mean that in a negative way. I mean it as I am now comfortable. I know that every day and even every flight has new wrinkles. While this flight may look like a repeat of yesterday's flight, it is not the same. It is tough to explain this, but I have found a way to understand it and think it may help elucidate what I am saying. To me, it is like building a house. You learn how to plan each critical task: the plumbing laid out according to the designer and where each wall will go; then you put up the frame based on that information. However, it is no longer like the house you built last because the switches are different, the inside walls are different colors, and the windows are different. It is easy to understand how this house will be put together because the basics are the same as the other houses.

For me, that comparison helps me know that I need to do, what I have done many times before, but this one will require different issues like the pattern you will fly. One of the most important tasks that's different is covering the other gunships break to the outbound. To protect it, terrain differences exist, and because of that, the enemy will be in different places. The bottom line is that it is the same as always, but different. Each time we shoot to eliminate the enemy, we may use a different ratio of miniguns to rockets. The terrain will be diverse even though it may look the same. Uneven ground may give the enemy grunts different places to hide? If the terrain is rockier, it may take more rockets to soften up the hiding positions, making the automatic machine guns more effective. Looking at it this way helps me make sure I recognize this target area. When I don't have to look at it the same as though I had never been there, I can spend more time looking for what is different; therefore, it is more accessible to try to locate the targets. I believe the system makes a person a better gun pilot and, hence, more valuable to the unit, which means you are better at protecting other people, whether in the air or on the ground. The fact that we go over virtually every flight down to every turn, inbound or outbound, makes us work better, too. Since flying guns is a team effort and working together, we can cover and protect ourselves and the ground forces. And that is the job we accepted.

I would be disingenuous to say we go into that depth of team debriefing after every flight. We don't. We're asked if we have any issues or ideas that we feel need to be discussed. In that hanger flying time, we listen and learn and talk and perhaps teach a different view of right and wrong. When I went to flight school, I felt that although we were working together as pilots, we still worked independently. As gun pilots, everything we did was done as a team which, when done correctly, made us much more effective. However, as gun pilots, our ability to work together made us much more deadly. The military uses the idea of symbiotic, or mutualistic, teams where both parties benefit from the interaction in most things. In this case, the team becomes much more effective because each aircraft is more effective.

THE SECOND MONSOON

It's the start of the winter monsoon season. My second after the summer monsoon season. Typhoon Hester has finally blown itself out after causing a lot of damage, wreckage, and mayhem, particularly in Chu Lai. I have seen heavy rainstorms in my life, but I've never experienced the deluges that occurred in the Central Highlands. Curtains of rain came whipping down in relentless waves, shrouding the whole world in a coffin of water. Between monsoons and Typhoon Hester, we were weather controlled, and flying was at a new skill level. On days I could fly, I had to work my way through the heavy rain shafts to be able to go to where the fight was.

I was thinking that if we counted every combat hour, we would find out we flew a lot of missions, sometimes several per day. It was not uncommon to leave an area we had been covering to move on to another battle. This situation could happen three or four times a day.

As aviation units scaled back preparing to stand down

and leave, the war went on. The North Vietnamese were not leaving this war. Fewer aircraft meant that we weren't even close to having the amount of aircraft coverage needed, so the demands for our service were hectic. Our men performed brave deeds, and yet, the awards were cut back because of the way the process worked. It took two witnesses to write the commendation: someone to write up a person and someone else who was not affiliated with that person. With this typical control in Army bureaucracy, not that many brave acts were seen by multiple people every day. The process directly impacted the sorrowful lack of awards our group received. We were still flying, and the fighting was building. I don't know that very many pilots were there working to earn awards, but I do know to some it meant a lot. For many commissioned officers, this was their way of gaining combat experience that mattered to their career track.

The Americans were leaving this war after more than ten years of training the South Vietnamese. It had been important to train them in ground and air operations, so they could fight for their continued independence. They were trained and equipped to continue this fight; however, they were fighting one of the most dedicated Communist-backed enemies in the long history of war.

Soon, two years after we left, the South Vietnamese lost the fight. Perhaps they lost their will to fight, but they could not out fight the North Vietnamese and their suppliers. The North Vietnamese had strong backing from both China and Russia. In the end, the South Vietnamese proved that they were no match for the North. Saigon would never be Saigon the capital again. Soon the former capital in the South would be called Ho Chi Minh City, and Hanoi in the North would be the capital of Communist-ruled Vietnam.

Since time began and man has walked on this earth, wars have been fought, most often driven by the desire for power and greed. Before his death in 1969 Ho Chi Minh had been instrumental in driving out the Japanese and the French, and history would credit his legacy as helping thwart all efforts to save the people of South Vietnam from the yoke of Communism. The intense fighting would escalate over the next few years as the people of South Vietnam would ultimately be consumed by the Communist machine. We fought for the people to hold on to their way of life. If anything is worth having, evil forces driven by greed and power will fight to take it away. In waging the war against evil, I can only say that I am proud to have given of myself, proud of the men who flew and fought, and proud of all those who didn't live to tell their stories.

ALL IN A DAY'S WORK

The month of November found me comfortable with my job, and with my firm belief that I was a good and capable combat pilot. I refused to cringe or jump when the enemy took shots at us, even on occasions when we took hits. Dangerous situations rarely, if ever, gave me cause to fear for my life or let go of my confidence. I was there to do a job, and I meant to do it. Fear did creep up on one occasion, but it had nothing to do with my personal safety. This variety of fear had everything to do with survival, however. It was a fear that my crew chief needed me to save his life; fear that I might not be able to do what needed to be done; fear that if I didn't do my best, he would die.

We were inbounding on a gun run, and my gunner and crew chief were hanging out the rear doors, shooting at any targets they could find. They were covering the inbound slicks and the lead gun on his break from outbound to inbound. These men were artists with the M-60 machine gun. Unbelievably, along with keeping the aircraft up, they cared for their M-60s with the utmost care needed, knowing that if things went south, these machine guns were our defense from NVA or VC chasing us. The M-60 fired roughly 400 rounds per minute but needed a barrel change every ten to fifteen minutes when under near-continual firing. When we were in the fight, these M-60s were exceptionally active.

I looked back, and the barrel on the crew chief's weapon was bright red. I saw him reach forward to the back of the AC's seat and pull back his hand. It looked like he was wearing a welder's glove. Then he grabbed that barrel, twisted it off, and put it in a bucket in front of him. He then grabbed another barrel that was hooked into the back of the pilot's seat, snapped it into position, and started shooting again.

Once that battle was over, after he returned to base and completed all the shutdown and post-flight items when I asked him what that barrel-change was about. He showed me how it was done and explained that it kept us in a full fighting position. Both he and the door gunner had three barrels to use when it was hot like that. I had noticed the barrels and had been curious, but I'd never see the action he performed enough to ask about the circumstances. He assured me that the welder's glove worked great.

At any rate, when we were inbounding on this gunrun, and the Crew Chief was hanging out the back covering lead, I heard him shout, "I'm hit! I'm hit!" I looked back quickly and saw blood on the floor, and he was half slumped over.

I called the lead and told him we were breaking off the run. I was heading over to the hospital at Cam Rahn Bay, and I had to be quick. The lead said he would cover as best he could and called for the backup gun to join in the fight. He knew it wouldn't take long for him to arrive there, and he knew I needed to get going. Luckily, we were not too far away from the hospital, and the Crew Chief had was still responsive, but I didn't know for how long. He said the pain was not bad. Then he added that it burned all the way up his back. In the chaos of battle, I anticipated the worst. Would he be paralyzed? Would the loss of blood kill him? Was I witnessing his last moments? Would his family reach out to me for details?

With all these thoughts going through my mind, I called the Tower. I told the operator that because I had a wounded crew member, we wanted the shortest path. I ended the call with, "We will stay low under any other traffic you have."

He was hesitant to clear me, so my next call was an emergency call. "Listen," I yelled, "The Crew Chief has been shot. I do not know what damage the aircraft may have if the round that hit him hit the aircraft. Finally, he cleared me to the pad directly ahead.

Then only a minute later, he emphatically bellowed, "You cannot land there if you are armed."

My immediate response, driven by my pissed-off position at his stupidity went something like, "You are putting protocol over saving the Crew Chief's life. You may not be aware, but there's a war out here, and we're a gunship. So, of course, we're armed. How do you think he was shot? We were in a battle."

He again refused my request. This time, I took a deep breath and simply said, "I have an emergency. I will use any spot on the airfield that I damn well please." Things went downhill for the next couple of minutes, but I saw the pad and executed an extremely fast, very steep bank to line up.

As I was losing air speed and altitude, I saw this jackass run onto the pad to stop me from landing, but I kept my nose high, rapidly slowing down. I had no intent to go around. This yahoo realized that I was about to put this aircraft on top of him. He moved aside, and I started the two-minute cool down to shut off the gunship. Then this idiot came running up to my door, trying to open it. My gunner jumped out and grabbed him to move him out of the way. The emergency medical team put my crew chief on a stretcher and whisked him indoors.

At the end of the two minutes, I shut it down and started getting ready to disembark the aircraft. I was eager to see who this clown was and why he was acting like this. He started screaming at me as the blades overhead were slowing down. I told him to shut up. At that point, I did not fully understand the full nature of his problem. He kept screaming over and over, "Remember flight school? Remember flight school?" Then he grabbed my arm and almost pushed me over. Now the game had changed.

This guy was an Air Force major and had about five inches of height and maybe sixty pounds on me. Once again, my gunner was grabbed and pushed away. I saw MPs running towards the aircraft. I finally got my arm free and reached for my pistol when almost everyone stopped except this guy. That allowed the MPs to see that he was physically attacking, so of them grabbed him and got him under control, telling this screaming ass to shut up, or they were going to leave him to my gunner and me. They escorted him out of the way as several medical people watched. I expected them to start on me, but as we went by, a couple of them thanked me for not putting up with him.

When we made it inside the hospital, we discovered that my crew chief would be released after some stitches and bandages. We were standing there when a couple of the nurses and a doctor asked to look at the aircraft. They had all been there for at least a year and had seen quite a few Huey's but not a gunship. Apparently, this major made this policy that no armed aircraft could land on this pad. I explained to them very succinctly that if one of ours was hurt, you would see us on this pad. He had escaped with his actions today, but never again. As we were ready to go, they forced this major inside, and the MPs agreed to watch him. Just another day at work.

The mission was over, so we went back to base, which meant that the crew chief could lie down and rest. He was fine, but he was going to hurt for the next week or so. As he was leaving the aircraft, he pulled me aside and apologized. He also thanked me for not delaying.

"We had a plan, which was to get you help as fast as we could. It wasn't a problem for me, but thank you for that," I explained.

"Well," he said. "I figure you might get in trouble later."

I replied, "I'm bound to get into trouble later, but probably not for this."

HOME AND BACK AGAIN

As life happens, time goes on. Even as I felt comfortable in my job and with my fellow gun pilots, we knew it was all going to come to an end. Even down-to-the-line pilot level, we knew the end was coming. and it was not going to be fun. We were keenly aware of the buildup in Cambodia and had heard the same about what was happening in Laos. We were finding NVA troops crossing the trail and farther into Vietnam. We had caught them in what appeared to be training situations, and that was not in their favor. The NVA had proven to be worthy opponents in many fights, but they had lost many combat troops in the time we American troops had been fighting in this war. They had to keep bringing in new troops who were not battle hardened. Everyone knew the Americans were preparing to leave, so the drive to takeover Vietnam was soon to happen.

This was not a time to be leaving if you believed in fighting to keep South Vietnam free from Communist rule, but the American people were war-weary and wanted us out of this war. As long as we were there, we could hold the enemy at bay, but the cost was high. As units stood down and returned to the U.S., the remaining troops were fighting an enemy who was growing in size while we were decreasing in number.

Very few large assaults occurred as in previous years, but the war went on. We did many short flights as we staged out in the field, so our flight times decreased markedly, but the number of battles stayed high. Many, or even most, days were busy with fighting with little time enroute.

As my time for R&R was almost up, I had to go now or lose the opportunity. I was left with the choice of going back to the States or no R&R. The plus side was that I'd take a one- week leave, and then I'd get two weeks off. I opted for that, and my time was coming up. I readied for my trip while we started gearing up for the unit to shut down. We had not been told directly that our unit was going very soon, but it was somehow hinted at, so our preparing for the inevitable was smart.

As I prepared to go on R&R, I looked around me and realized I had turned into one of those war veterans who realized the bond between us, my fellow soldiers I had grown to know and to understand on a visceral level. I would later hear the saying "We went over as strangers but returned as brothers." We literally kept each other alive but also sane, or at least close to sane. We talked whenever we had some quiet time, and we learned of each other's lives and families, including girlfriends and wives. Finally, we felt we knew them, also.

With all the violence and hatred in a war, there had to be some sense of sanity, and we provided that for each other. That is not to say we were all singing songs of love and caring. We had our disagreements, and some we did not like much, but even we had respect. Once I sat down and thought things out about who and what we were, I acknowledged the skill and bravery that was demanded. I may not like the guy I was with, but I'd fly through the gates of hell to bring him back if it was no more than impossible.

When it reached undoable, I left a little bit of my soul with them. I could not forget them, and as I would find out in later years, we never would. I did not want to leave and no longer be a part of these soldiers' lives, but I also knew I needed a break. To be honest, I had believed from the beginning when I volunteered that I would not return alive. I do not know why I felt that, but to me it was real. Now that I had survived and done well at my job, I was beginning to think that maybe, just maybe, an outside chance could mean that I would do the job right and make it through this. I also knew I needed a change of scenery, even if for only a week or two.

So, I got ready and went through that long, long trip home. I understood why they gave us the extra time off. Going through the international date line and through roughly eleven or twelve time zones between Texas and Vietnam and back was a little hard on my mind and body. I was positive on my return that I would be right back into the war.

Soon after arrival, it happened for the first time; I was called a "baby killer" by some long- haired hippy freak. I was spit on by a couple of jackasses who hadn't been man enough to join the war effort and belittled those who did. Each time they had numbers on their side, but I was prepared to fight. Things didn't escalate, and I stopped short of being hurt badly by them. Actually, they didn't want the fight, just the opportunity to run their mouth. I was disillusioned and unprepared for these occurrences, and it hit hard. Going back to Viet Nam, I felt I was rejoining my own kind.

As I suspected I was going back to work when we found out that we were standing down and the unit was going home. I was told not to unpack as I was leaving right away with credit for a full tour because I had already served over six months. The next morning, I found out I was going to Pleiku to join a cavalry unit as a scout pilot. This unit had suffered so many losses they were pulling pilots that had been trained in the cavalry aero scouts' role or as a gun pilot, and as it was, I was both. My platoon leader was going with me.

I tried to explain that I had been told I was going home, and Pleiku was certainly not my home. We were picked up in the morning by one of our slicks and flown up country to that base. It was a hard farewell to my buddies and the others for me to head into this part of the war. We had heard stories all through my tour about this cavalry unit and their bravery but with high combat losses. I was bordering into the onset of depression until I caught myself and talked myself into a better frame of mind for this transition. I would soon find out what it was like in this type of unit.

* * *

As we stood on the flight line for the last time at our home base, I looked closely at the aircraft. Soon to be my last company I flew in, I realized how much I was going to miss it here. We faced the enemy very successfully for the most part. I had to say goodbye to my aircraft that had worked so well and reliably when we were facing the enemy.

I had grown from a nineteen-year-old new guy to a twenty-year old, battle-hardened combat pilot I was today. Viet Nam would always be where I grew up. I would remember the experience tied to a great first combat unit, and I suspected I would need all I had learned and more.

As for my fellow pilots in the 92nd Assault Helicopter Company, I will always remember them as the outstanding pilots they were. I had made some friends, but war has a way of taking them away. Most I would never see in person again, but I will always consider them my brothers. Leaving was hard, but the needs of the Army outweighed any desire I had to stay, so I had no choice but to "soldier on." I vowed I would do the best I could. "Parting is such sweet sorrow" has a new understanding for me, and it was time to move to the next place in the war. There are stories to tell from there.

LESSONS LEARNED

1. I learned that so much rain could change your thinking. As I became increasingly tired of rain, I started thinking that I might survive it. Hoping that we didn't have terrible weather only produced more relentless rain.

2. I learned to wonder how many ways there are to play blackjack.

3. I learned that rain meant no entertainment and that it rain could stop someone from singing. (I already knew the rain kept the singers from traveling.)

4. I thought I learned rain must have slowed time down, but I don't think that is the case.

5. I learned that the bad guys had taken the rain time to restaff their units, so the action got hotter. No way to win this.

6. In the beginning I had the internal conflict that I should not kill, but I also believed I should help people who are being harmed when they cannot protect themselves.

7. Along with that, having had the life of travelling to foreign countries and living in some of them not long after World War Two had ended, I understood what war can do to countries and to people.

8. I learned what U.S. military had done to stop the militaries that were causing the wars, and I came to understand war at least to some extent.

9. I learned I had not learned anything about the horrors of war. I had always been proud to be an American and proud of those, like my father, who served and fought in both Korea and Vietnam, serving their country at the risk of their own life. Now I felt we had a whole new generation of Americans that served, and I was proud to be a part of that.

ADDENDUM

No event in American History is more misunderstood than the Vietnam War. It was misreported then, and it is misremembered now. Rarely have so many people been so wrong about so much. Never have the consequences of their misunderstanding been so tragic.

President Richard Nixon, 1985

QUESTIONS ABOUT KILLING: AN EXPLANATION

One of the more unsettling things about being a combat veteran, and I believe this is perhaps more than normal than unsettling, is the question: "Did you ever kill anyone in the war?" That question is sometimes followed quickly with a barrage: "How many people did you kill? What weapons did you use? How did you feel about killing them?" These questions asked by boys or teenagers are understandable because they may never have had the opportunity to ask a war veteran such questions. Yet, in today's world of video games, killing is a norm and, therefore, those questions might not seem like weird questions at all. At the end of the Vietnam War video games were not out, and television shows were more restrictive than today, so at the time, the boys who asked me those questions were just trying to understand. When adults ask these questions, I want to answer, and have a couple of times, with the affirmative followed by "How about you? Did you ever kill anyone?" Of course, that is not politically correct, and possibly, the question may just be coming from someone who, like the boys, hasn't interacted with true combat veterans. At this point a distinction must be made. Many service members are serving in a war zone, but they are not combatants or combat veterans. At some point, the distinction was pointed out that for every warfighter there are seven people serving in the war zone but not actually participating in the fighting. That is not to say the dangers of incoming rockets, mortars, or sappers is not the same level of deadly as the actual combatant.

When the question of killing is asked, several things may be occurring in the veteran's head. One may be a gut-wrenching fear of saying the wrong thing. These kinds of questions also put Vietnam veterans out of the mainstream even more than before people knew that they were in the "kill or be killed" group. He will be separated even in the group when people know the answer to those questions. This is on top of the fact that the veteran may have already realized they *are* different from others. When a young combatant comes home, they are not the same. What they have to talk about in the time from when they left for the service and the time they returned from the war, they will have almost nothing to talk about with the people who were their friends. I was once asked why I was so quiet when I was outgoing and friendly before. My best answer was, "I didn't do the same things you did, and you did not see what I saw and did." Even if you think it is a bit of bad taste for people to ask questions, you probably should not say so.

It sounds simplistic to say, but war changes you. With some observation, people can probably tell the difference. Possibly, you will be quieter and more reflective. You will most likely sit in public places like restaurants with your back to a wall, so you see everyone coming in and leaving and where they sit. More importantly, some circumstances may bring back in vivid detail one or many kills. There is killing in war. The duties the veteran performed could indicate the level of killing that they may have been involved in.

You may also encounter the person who wants the down and dirty details such as, "What weapon did you use? What did the enemy do when you were eliminating him? How did you feel?" This is down to the nitty gritty of the act. You can answer any way you want, but most veterans understand this is a serious question that deserves a serious answer.

As a gunship pilot, I am always amazed that once people knew what I did in the war, the question of killing seems odd since your job was to shoot and fight as hard and accurately as you could, so killing *is* the job. That does not imply reckless or wild shooting with no regard for the peripheral danger to non-participating people.

For some, the killing did not seem to bother them as much. Of course, a calm outer manner does not indicate the inner person. If asked if a veteran remembers specific battles or events in a battle, I believe the answer will be unanimous "yes." The memories won't be   fond memories although to be quite honest, if you had a particularly hard target due to location on the ground, or you see the enemy inflicting heavy damage on friendlies or even your own aircraft, you may remember that with some satisfaction. I believe we all have some of those.

The other thing about this killing issue is when someone says, at the end of your comments, "I could never do that," All I can figure out to say is, "Well, I understand that, but I personally believe that if someone close to you, your own family or friends, is in mortal danger and you have the tools, you probably could do it. That is not to imply you would not have a difficult time dealing with the event."

Soldiers are people, too, and because of that, serious issues exist. When dealing with my PTSD and looking at this subject, I always realize that this may be the hardest subject I have to work out. What I consider a strange quirk in my personality allowed me to do my job as a gunship pilot. I prefer the job

Military Occupational Specialty (MOS), which was attack helicopter pilot. I feel this is more easily understood by those who are not familiar with titles used by the Army, but it was and still is I believe 100EO.

I had, after my first Kill, no reservations shooting mini guns with the intent of eliminating them from the battlefield, or just saying killing them. It took almost no time to see what the NVA and VC did to South Vietnamese rice farmers and their families. It became very clear to me that they had no reason to be left alive to continue what they had done. The elections were coming up and the enemy knew that they would not win if the people had a free and fair election.

With the U.S. there and United Nations monitors to come in, they had to make the people vote the way they wanted them to. By terrorizing the people of South Vietnam to this level, the people were afraid to vote as they wished to.

What made things even worse was the fact that China and North Vietnam were paying these people who were committing heinous war crimes bonuses for committing atrocities like beheading men, women, and children. They would remove the bread winners so that the family could barely make it after the attack. When we were sent into battle, we understood that these soldiers had no moral compass. I had no compassion for them. Death was what it took to make them stop, so death for them it would be. It was no problem for me to do the job I was tasked to do. At night, I often tried to understand why we were now judges, jury, and executioners. The truth was easy enough: leave this country you have invaded or that you had turned against, leave and we would not hunt you down. Stay here in an illegal invasion and if we could get to you, you would die.

In general, war is hard to reconcile given what people can do: wonderful things such as health care, freedom of religion, the right for free and safe elections, helping one another and protecting animals. It was never easy to deal with these issues but at least we were trying; whereas the enemy was not.

I am at peace for the most part of this, but, that being said, I still deal with some things, and I always will.

WHAT THE SOUTH VIETNAMESE WANTED

As I understood it, the South Vietnamese people wanted something simple; they wanted the North Vietnamese out of their country. Although the history of war after the fact is very messy business, from my perspective, they wanted the right to choose their own form of government and the right to be left alone. In other words, what they wanted has historically proven to be the exact opposite of Communism. Ho Chi Minh, the self-appointed dictator of North Vietnam, had joined forces with the U.S.S.R, China, and North Vietnam to spread Communism to South Viet Nam. Ho Chi Minh's heavy-handed leadership had been honed during World War II when he declared Viet Nam's independence from France shortly after Japan surrendered. After the French were routed at the Battle of Dien Bien Phu, Ho Chi Minh started the takeover of South Vietnam with the sole object of reuniting North and South Vietnam with him in place to lead this now Communist nation.

It was in 1954 that the Geneva Accords split Viet Nam at the 17th parallel. The hope was that Viet Nam would eventually be united, and various prospects attracted the attention of the world's superpowers, those on the side of democracy and those who opposed it.

The South Vietnamese had elected many poor leaders into powerful positions through ignorance of the truth about the people they elected. These bumbling bureaucrats had drained the money out of Vietnam's coffers. The country was about to collapse into even further chaos, having been through twelve weak governments installed between 1963 and 1965.

The Viet Cong emerged to present themselves as "the solution" to removing these corrupt politicians with the promise of replacing them with good leaders. Of course, they were only good leaders if you supported the tenets of Communism which they supported.

The country was on the brink of civil war when the Communists increased their strength, helped the VC, and saw an opportunity to expand power and take over the South. The possibility of war was a reality. It was during the height of the Cold War when President Eisenhower forwarded the Domino Theory that a Communist victory in Viet Nam would create instability and the spread of Communism throughout South Asia.

As time passed, one incident occurred between a U.S. ship and a Russian-made North Vietnamese vessel. Shots were fired and President Lyndon Johnson, who became President after John F. Kennedy's assassination, stepped up to the challenge. Then on August 7, 1965, Congress passed the Gulf of Tonkin Resolution, granting Johnson the power to "take all necessary measures to repeal any armed attack against the forces of the United States and to prevent any further aggression." The war and our part in it had a clear rationale.

It was in January 1973, when the Paris Peace Accords

attempted to create terms of a fragile peace. The terms of the Accord were straightforward. They included a cease-fire agreement, the departure of U.S. forces, the release of prisoners of war, and the eventual unification of Vietnam through peaceful means. The last of the American troops would be leaving soon.

The North broke the terms of the agreement and Saigon fell to Communist forces on April 30, 1975. As always happens when Communists take over, malcontents are sent to reeducation camps and forced to follow the Communist doctrine or be killed.

It sounds horrible and I know it must have been; however, the South Vietnamese had been thoroughly shown the Western non-communist way of life, but their forces were no match for the strength of the North, no matter the cost of losing. After the fall of Saigon, when he signed the agreement to end the war and accept the surrender of the South, North Vietnamese Colonel Bui Tin reportedly said, "You have nothing to fear; between Vietnamese there are no victors and no vanquished. Only the Americans have been defeated." To this day in Viet Nam the war is referred to as the "American War."

After more than forty years of Communism the South Vietnamese have created a market economy and some freedoms are now coming home to these people. It is not over yet, but right now it looks like the semblance of a free way of life is working; however, tourists are warned not to discuss politics and that any criticism of the government can result in extremely harsh punishment. There's little freedom of the press; no real freedom of speech, and government control is enacted regarding many of the privileges that we take for granted.

VIETNAM WAR FACTS, STATS, AND MYTHS

I recommend the following websites if you want to research pertinent information that will help you understand the causes, battles, accurate statistics, and reasons for American involvement in the Viet Nam War. Regardless of news coverage, then and now, I was there. I witnessed events first-hand. I fought alongside our allies. In the years that I was in Viet Nam, I flew more helicopter hours in battle than a civilian pilot would fly in a lifetime. You might think you know the truth, but I'm here to tell you: perceptions can be manipulated by power.

US Wings, Established 1986. Online since 1994
https://www.uswings.com

This comprehensive website includes detailed casualty statistics and a long series of unsubstantiated common beliefs or "myths" with full well-researched rebuttals provided for each one. An online store with military clothing and other accessories is included, also.

Vietnam War Bibliography
https://edmoise.sites.clemson.edu/bibliography.html

This online listing of more than four thousand references was created by Professor Edwin Moïse, Clemson University. Copyright © 1996, 2002, 2003, 2005, Edwin E. Moïse

Viet Nam War 50th Year Commemoration
https://www.army.mil/vietnamwar/history.html

Sponsored by the U.S. Army, this comprehensive website moves through a chronology of the Viet Nam War with pictures, dates, and highlighted information and overviews about major offensives and objectives